Superbugs

Superbugs

DEADLY MICROBES
AND THE EXTRAORDINARY
RACE FOR A CURE:
A TALE OF HUMAN INGENUITY

MATT McCARTHY, MD

AVERY

an imprint of Penguin Random House

New York

AVERY

an imprint of Penguin Random House LLC
penguinrandomhouse.com

Most Avery books are available at special quantity discounts for bulk purchase for
sales promotions, premiums, fund-raising, and educational needs.
Special books or book excerpts also can be created to fit specific needs.
For details, write SpecialMarkets@penguinrandomhouse.com.

LIBRARY OF CONGRESS CATALOGING-IN-PUBLICATION DATA

Names: McCarthy, Matt, author.
Title: Superbugs : the race to stop an epidemic / Matt McCarthy.
Description: New York, New York : Avery, [2019] | Includes bibliographical
references and index.
Identifiers: LCCN 2018050075 | ISBN 9780735217508 (hardback) |
ISBN 9780735217522 (ebook)
Subjects: LCSH: Drug resistance in microorganisms. | Antibiotics—Research. |
Bacteria. | BISAC: MEDICAL / Research. | MEDICAL / Clinical
Medicine. | MEDICAL / History.
Classification: LCC QR177 .M33 2019 | DDC 616.9/041—dc23
LC record available at https://lccn.loc.gov/2018050075

ISBN 9780735217515 (paperback)

Printed in the United States of America
1 3 5 7 9 10 8 6 4 2

One sometimes finds what one

is not looking for.

—ALEXANDER FLEMING

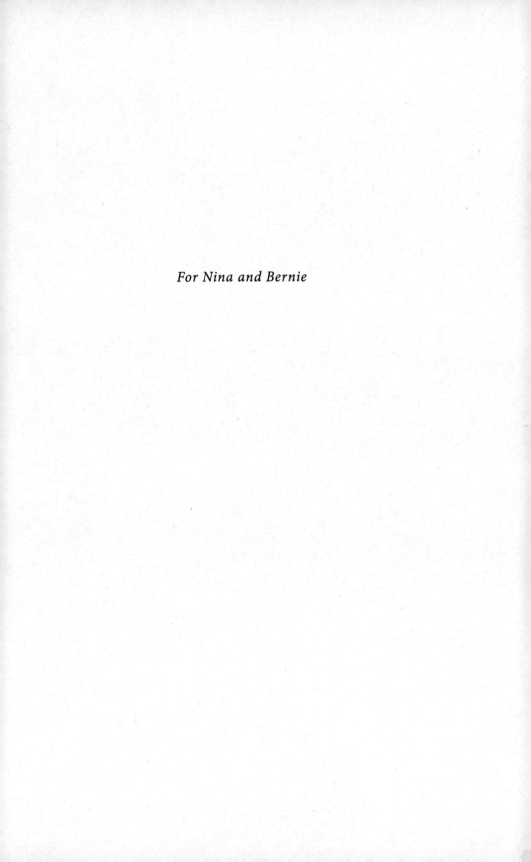

For Nina and Bernie

Contents

Author's Note

THIS IS A true story about a clinical trial, and the people I have written about are real. However, in order to ensure patient privacy and maintain the confidentiality of others, this work has been carefully vetted to comply with the Health Insurance Portability and Accountability Act (HIPAA), and throughout the book, names, dates, and personal identifying details have been changed.

Superbugs

Prologue

IT WAS JUST after dawn when I felt the buzz on my hip. I broke stride, put down my coffee, and glanced at my pager: I was needed in the emergency room. It was 2014, an unseasonably warm October day, and the text induced a flurry of anxiety and excitement. After eleven years of training, I had accepted a position as a staff physician at NewYork-Presbyterian Hospital, a tertiary care center on the Upper East Side of Manhattan, and a patient had just arrived with a perplexing infection, one that had stumped the team in the ER.

A moment later, I was standing before a group of medical students and residents and my new patient. The young man writhing on the stretcher was an African American mechanic from Queens named Jackson, with dark-green eyes and a small Maltese cross tattooed onto his neck. He had been shot, and a large area surrounding the bullet, which was still lodged in his left leg, looked infected. As I peered into jagged edges of the entry wound just above Jackson's knee, a student handed me a piece of paper. The printout revealed the results of microbiological test, which caused my eyes to bulge. My patient, I discovered, was infected with a nimble and aggressive new bacterium that was resistant to every antibiotic at my disposal, except for one: colistin.

I had used the drug only a few times in my career and never with good results because it was so outrageously toxic. Colistin might kill bacteria, but it destroyed kidneys and other internal organs in the process, leaving many of my patients with just two options: dialysis or death. Antibiotics that had proven so effective just a short time ago were now useless, and if I wanted to save this young man's leg, it was my only option. I shook my head and handed the paper back to my student. "Not good." More than twenty thousand people die every year in the United States from antibiotic-resistant infections, and the pipeline of drugs to treat them is always on the verge of drying up. I crouched to meet Jackson's eyes and carefully considered my words. "You have an infection," I said. "A severe infection."

The man's gaze darted from me to the men and women standing in a horseshoe behind me. "How severe?" He took in a small breath of air and held it, waiting for me to say something. It felt like an hourglass had been flipped; suddenly the tiny room was very hot. I took off my white coat and rolled up my sleeves. "Quite severe."

His eyebrows raised, and I reflexively extended my arm to hold his hand, but caught myself. I wasn't supposed to touch this patient without protection. I pivoted back to my team. "Everybody out. Now." I pointed toward the door. "I'll be right back." Just outside of his room, I put on a disposable yellow gown and a pair of purple nitrile gloves, and returned to the bedside alone. "It's very hard to treat," I said, "but not impossible."

Jackson was now breathing very quickly, on the verge of hyperventilating, as sweat beaded on his forehead. He grasped his thigh, inches above where the bullet had entered. Beneath his fingertips, bacteria were rapidly multiplying, devouring muscle and bone. "Am I gonna lose it?" he asked. "The leg?"

In truth, I wasn't sure. Only colistin had a chance of destroying the infection, but there were no guarantees. The last person I prescribed it to died twelve hours after she received it. The one before that died while receiving it. "I don't think so," I said, as confidently as I could. I squeezed his sweaty hand and tried to imagine how I would summarize the nuances of the case for his wife and children. They would need to take special precautions just to be in the same room with him. "We're going to get through this," I said as his eyes began to water. "We will."

I left the room, removed my gown and gloves, and addressed my team. "Start colistin," I said. One of the residents frowned as she scurried to a computer to put in the order. Then we vigorously washed our hands and moved on to the next patient.

When rounds were over, I walked across the hospital to the office of my research collaborator, Tom Walsh, director of the Transplantation-Oncology Infectious Diseases Program. Walsh is a wisp of a man, pale and thin like a potato chip, with deep-set eyes, a warm smile, and a surprisingly firm handshake. His modest features are a notable contrast with my own: I have a high forehead, broad shoulders, and a nose that's slightly too large for my face.

We make for an odd pair.

Walsh is one of the world's leading authorities on obscure infections, and when he's not caring for patients, he's creating new antibiotics to treat them. We had met a few years after I graduated from medical school—I still have the elegant biochemical structures he drew for me during our first interaction—and I've been working with him ever since.

In 2009, he moved from the National Institutes of Health (NIH), the federal agency responsible for biomedical research and disease prevention. Walsh brought with him an expansive research consortium—an international team of physicians and scientists who conduct experiments in test tubes, animals, and humans—to develop antibiotics. He is one of the only researchers in the world to oversee a laboratory of this scope; he is an expert in infectious diseases, oncology, pediatrics, internal medicine, pathology, microbiology, and mycology. No one else possesses his breadth of knowledge. Not surprisingly, Big Pharma is eager to work with him. But Tom Walsh does so on his own terms; I once saw him quash a $50 million drug development initiative with three barely audible words: "Would not pursue."

He had called me that October morning in a fit of excitement, with news that Allergan, the pharmaceutical giant, wanted us to run a clinical trial: a large-scale human experiment with an unproven drug. The

Dublin-based company was developing a promising new molecule and it wanted us to show it was not only safe but effective in treating humans infected with antibiotic-resistant bacteria, known colloquially as superbugs. They had become a persistent problem for us; superbugs didn't really exist before the 1960s, and they were only sporadically seen in the world until the 1990s. But a combination of poor prescribing practices by doctors along with the indiscriminate use of antibiotics in commercial agriculture and farming exposed bacteria to our precious arsenal of effective drugs, and the microbes figured out ways to neutralize them. Superbugs were now everywhere—even on stray bullets in Queens—and they had become a leading cause of deadly infections in humans. "So, what is it?" I asked Walsh as I entered his office. He leapt up from his messy desk, hurrying past framed diplomas and awards that covered every inch of the mahogany walls, to greet me. "What's the drug?"

Walsh looked exhausted—the man regularly slept only three hours a night—because we were in crisis mode, desperately searching for new antibiotics to treat our patients. I had grown accustomed to watching men and women succumb to infections that had been treatable just a few years ago. When Walsh shook my hand, he brightened. "Dalbavancin," he said softly.

My fingers and wrists were still damp from the tense exchange in the emergency room; I wiped them on my khaki pants and sat down in the chair next to his desk. "You're kidding."

He handed me a thick manila folder. "I'm not."

Just the word—*dalbavancin*—brought me back fourteen years, to my days as an undergraduate tinkering around in the laboratory of a future Nobel laureate named Tom Steitz, a biophysicist who was known around campus as "the Michael Jordan of crystallography," the branch of science that probes the atomic building blocks of life. Steitz studied protein synthesis, an essential function of nearly all living things, and his discoveries led to all sorts of new drugs, including a handful of antibiotics related to dalbavancin, called "dalba" for short. Like Tom Walsh, he was a visionary who could see drug development in ways that others couldn't.

I connected with Dr. Steitz through his son, Jon, who happened to be my teammate on the Yale University baseball team. He and I were pitchers and biochemistry majors, and we were both drafted out of college to play professional baseball; Jon was selected by the Milwaukee Brewers in the third round of the 2001 Major League Baseball draft, and I was taken the following year, in the twenty-first round, by the Anaheim Angels. We briefly thought we were destined for the big leagues.

A year later, after a stint playing minor league baseball in Provo, Utah, I was cut by the Angels and exchanged my baseball mitt for a stethoscope. I enrolled at Harvard Medical School in the fall of 2003, moving to Boston around the time Jon gave up the game and went to Yale Law School. A few weeks after classes began, I attended a lecture by a young and charismatic infectious disease doctor named Paul Farmer, cofounder of the global nonprofit Partners in Health, and immediately knew what I wanted to do with the rest of my life. I was going to study infections to learn how to defeat them.

"Let's get to work," Walsh said, snapping me out of my reverie.

This was the moment everything changed, when I went from a passive observer of drug resistance to an active participant in the race to stop the expanding threat of superbugs. But before I could start the long and winding journey of a clinical trial, I had to familiarize myself with the painful lessons learned from generations of failed studies and appalling ethical lapses, as well as the remarkable scientific advances behind the work of Tom Steitz, Tom Walsh, and others. That extraordinary story, the one that ultimately led me into the hospital room of a terrified mechanic from Queens, begins with a different bullet wound one hundred years earlier, in October 1914, when a soft-spoken military physician noticed something unusual and had a hunch. It's an adventure dotted with clues that would help me unravel the mystery of Jackson's infection.

PART 1

A Chance Observation

CHAPTER 1

The Fog of War

THE DOCTOR EXAMINED the fresh wound and shook his head. The bullet had pierced the soldier's right thigh, pulverizing his femur before exiting the back of the leg, leaving behind a bloody mess. It would soon become infected, and the physician, a captain in England's Royal Army Medical Corps, closed his large blue eyes and imagined what was to come. There was no shortage of terrible fates befalling soldiers with this kind of injury, from amputation to gangrene, even organ failure. But he was most worried about tetanus—a lethal condition causing paralysis and eventual suffocation—that was terrorizing so many British soldiers in his battlefield hospital on the Western Front.

It was October 24, 1914, and Alexander Fleming, a thirty-four-year-old doctor from Scotland, was caring for a throng of men at a makeshift military base in Boulogne, France, that doubled as a wound-research laboratory. The Great War was just eleven weeks old, and already losses were heavy. British soldiers had arrived in France on August 7; two weeks later, French and British infantry were brutally defeated by the Imperial German Army in a forest fight at the Battle of Ardennes. The unexpected drubbing had triggered a slow and humiliating retreat as the Germans continued their march toward the French capital.

Then, on September 6, something astonishing had happened: thirty miles northeast of Paris, six French field armies and the British Expeditionary Force suddenly halted and counterattacked. For three gruesome days, the battle shifted back and forth along a one-hundred-mile front. Owing to remarkable advances in artillery—powerful new machine guns, howitzers, and mortars—it was one of the bloodiest engagements in the history of warfare.

The Allies' gambit worked. Devastating losses forced the Germans to abandon plans to invade Paris. The victory, however, came at an extraordinary cost: more than two hundred thousand French and British soldiers were wounded at what would become known as the first Battle of the Marne. In its aftermath, waves of injured men, bathed in blood and riddled with shrapnel, were brought to Fleming's hospital.

The young doctor grabbed a damp cloth from his surgical bucket and dabbed the soldier's leg, cleaning mud, blood clots, and shreds of uniform from the gaping exit wound. He picked up a scalpel and carefully excised a small swatch of fabric from the man's muddy pant leg. This piece of clothing, Fleming hoped, would solve one of the most puzzling questions of World War I: Why were so many soldiers dying of tetanus?

It was a rare disease, typically infecting only one in every hundred thousand people, but in Boulogne it was everywhere. Fleming suspected that the bacteria causing tetanus was embedded in British military uniforms. When soldiers were shot, he reasoned, the organism was introduced into the bloodstream, overwhelming the body's defenses. Fleming rushed to his laboratory, handling the scrap of dirty clothing with great care. He passed row after row of camp beds holding injured soldiers from Marne, and Mons and Ypres in Belgium. Many had lain on the battlefields for days.

The makeshift laboratory was located in the musty basement of an old casino, beneath high-ceilinged, ornate, once-elegant rooms, and it was covered with signs of Fleming's ingenuity: incubators he'd heated with paraffin stoves, Bunsen burners running on alcohol, glass-blowing burners using fire bellows, and a matrix of petrol cans and pumps to supply water. At his lab bench, Fleming squeezed the small piece of fabric into an empty test tube and placed it next to a row of tubes that were all incubating clothing from injured soldiers. After adding a special broth to the glass, Fleming returned to his patient and went about the task of dousing the man's thigh with antiseptic fluid.

Peering closely into the wound, Fleming could imagine how this would play out: For the next few days, discharge from the leg would be reddish-brown and foul smelling, consisting mostly of clotted blood and bacteria. After a week, the dank material would lose its pigment and odor, gradually transforming into a thick soup of pus. If the soldier was unlucky, as many Brits in France were, he would develop fever, restlessness, irritability, palpitations, and, finally, the telltale sign of tetanus: lockjaw. Tetanus caused facial spasms that left many soldiers with a permanent grin—a perverse condition called risus sardonicus—before inducing paralysis and an agonizing, slow death.

The bacterium responsible was a common inhabitant of horse intestines, and its spores—the reproductive units—could lie dormant for years in soil containing traces of manure. Tetanus was an anaerobic organism, which meant it did not grow in the presence of oxygen. Even brief exposure was enough to kill it. So why was it flourishing on the well-tilled fields of Belgium's Flanders region, where oxygen exposure was constant? And, more importantly, in wounds exposed to air? Fleming thought that the bacteria were hiding out under shrapnel, within the recesses of the wound, where oxygen was scarce and antiseptics were washed away by the discharge of pus. That was why the harsh chemicals that easily killed tetanus in a test tube failed to do so in flesh.

Fleming had come to France at the behest of his mentor, Almroth Wright, a controversial figure who had been the first to mass-produce a vaccine against typhoid fever. In contrast with Fleming, who, owing to his small stature, was often asked to play women in his dramatic society—he took on the role of a vivacious French widow in a production of Arthur W. Pinero's comedy *The Rocket*—Wright was a bear of a man, with a bushy brown mustache, small spectacles, and lock of wavy hair parted fiercely to the right. Some suspected he suffered from a hormonal disorder. They, too, made for an unusual pair.

Wright had lobbied hard for his typhoid vaccine to be given to British troops during the war, and when there was some initial hesitation—routine vaccination was not yet *en vogue*—Wright published an impassioned plea in the *London Times*, "On the Inoculation of Troops Against Typhoid Fever and Septic Infection." The piece had appeared four weeks earlier, on September 28, 1914, just seven weeks after Britain had declared war on Germany in response to its invasion of France. Although the public appeal was unpopular with many doctors—some of whom referred to him as "Almost Wright"—it worked, and the British army quickly made vaccination against typhoid compulsory.

Sir Almroth Wright

Wright had also recommended vaccination against sepsis, but the director general of the British Army Medical Service, Sir Alfred Keogh, was unconvinced. He suggested more research was needed before mandating a second inoculation. Wright created a wartime research unit to study the bacteriology of wound infection, and this was where Fleming now found himself.

Surrounded by infection, unable to help the thousands of men who were suffering and dying around him, Fleming had become consumed by a desire to discover something that would save his soldiers. But at the moment, he had only antiseptic fluid, wound dressings, an untested antitoxin generated from horse blood, and his scalpel, none of which could protect them from a bacterium that was proving remarkably hard to kill. For some of his infected men, the cure would involve a hacksaw.

The medical world in which Alexander Fleming toiled as he shuttled between injured soldiers and his casino laboratory was defined by two approaches to the treatment of infected wounds: the physiological school, which concentrated its efforts in aiding the natural protective agencies of the body against infection, and the antiseptic school, which aimed at killing the microbes in the wound with a chemical agent.

Fleming knew antiseptics worked well in theory, but he worried that the active ingredients—caustic chemicals such as boric acid, flavine, and carbolic acid—might actually be harming his patients. The soldiers simply didn't get better with antiseptics, and the doctor had a hunch that they might in fact allow the tetanus bacterium to proliferate.

His theory was that abrasive chemicals might work in the central cavity of the wound, but they lacked the penetrative power to clean the tissue at the edges. Something about the periphery allowed bacteria to thrive. It was a radical thought, one that would have had Fleming laughed out of any hospital in Europe, but he was increasingly certain that antiseptics were killing his men, and he was designing an elegant experiment to prove it—one that drew on his life before medicine.

Prior to enrolling at St. Mary's Hospital Medical School in Central London in 1903, Alexander Fleming had learned an odd craft: glassblowing. Mostly he made tiny figurines—glass cats and scampering mice for family and friends—but when resources were scarce, he made his own research equipment, including test tubes. In Boulogne, Fleming began dreaming of ways to design tubes with contours that would approximate the jagged dimensions and texture of a bullet wound. The experiment was still in its infancy, but if it worked, it would turn the treatment of combat wounds on its head. Antiseptics were central to medical care during the Great War; British military policy mandated their use. Fleming was convinced they were not simply useless, they were dangerous.

But Little Flem, as he was known, was not drawn to controversy, or to combat, or even to conversation. (One colleague claimed that trying to speak with him was like playing tennis with a man who, when he received a serve, put the ball in his pocket.) The doctor knew he had a story to tell; he just had to write it.

MORE THAN SEVENTEEN MILLION military personnel died during World War I, many of them from tetanus. After the fighting was over, Fleming returned to London and to his lab bench in the Inoculation

Department at St. Mary's Hospital. By the time the armistice was signed on November 11, 1918, Fleming had published a dozen papers based on his work at Boulogne, and he was known in academic circles for his ingenious experiments with glassware. But he was a lone voice, and antiseptics still ruled the day.

Haunted by what he had seen on the Western Front, the young doctor spent the next decade in his laboratory, just up Praed Street from Paddington Station, trying to devise ways to destroy harmful bacteria and improve the treatment of infections. It was tedious work, staring at thousands of bacterial colonies in petri dishes in a dimly lit laboratory day after day, but it fulfilled him. He was consumed by a desire to understand how bacteria thrived and, more importantly, how they could be killed.

A chance observation in September 1928, a decade after the war, was briefly cause for celebration. One afternoon, Fleming noticed that the *Staphylococcus* bacterium—one of the pesky organisms that was so prevalent in battlefield wounds—was killed in the presence of a fungus called *Penicillium rubrum*. This accidental finding occurred in a discarded petri dish and led to the discovery of what he called a slow-acting antiseptic. Fleming dubbed it penicillin.

On May 10, 1929, he sent his findings to the *British Journal of Experimental Pathology*. Fleming wrote: "Penicillin . . . appears to have some advantages over the well-known chemical antiseptics. . . . If applied, therefore, on a dressing, it will still be effective even when diluted 800 times, which is more than can be said of the chemical antiseptics in use." The utility of this finding, however, was not yet clear. Penicillin could kill bacteria in petri dishes and test tubes, but it failed in the presence of blood. Because the fungus took several hours to exert its effect, Fleming was resigned to the fact that while penicillin might work superficially, it would be destroyed in the human body before it could ever kill the bacteria in a festering wound. Penicillin wouldn't save injured soldiers or anyone else. Instead, he thought it would serve as a valuable tool for preventing Staph bacteria from contaminating laboratory experiments.

Fleming was not the first scientist to notice that microorganisms could kill bacteria. Others had similarly suspected that their fungal extracts were either too feeble or too toxic to treat human bacterial infections, and they were discarded into the dustbin of history. They simply didn't realize they were on the precipice of something that would alter the course of human health forever.

Regrettably, Fleming's penicillin paper was not written in a way to make his findings accessible or reproducible. It was as if the manuscript had been dashed off in midthought. He did not explain how the molecule was purified from the fungus or where one might gain access to his chemical reagents to replicate his work. And he was such a poor public speaker that his lectures did little to inspire colleagues. To make matters worse, Fleming's collaborator had misidentified his fungus: it was *Penicillium notatum*, not *Penicillium rubrum*. Anyone hoping to reproduce his experiment was out of luck.

Investigators at Oxford University and Sheffield University Medical School, however, agreed with Fleming's assessment that because it killed off laboratory contaminants, penicillin could be useful to isolate and study a bacterium called *Bacillus influenzae* (now called *Haemophilus influenzae*), which some thought was responsible for the influenza pandemic of 1918. The outbreak had begun in Spain in May of that year, as World War I was coming to a close, and toward the end of Fleming's deployment, cases of influenza at his French hospital far outnumbered the wounded. By 1919, twenty million people had died from the infection, and the urgent need for some understanding kept Fleming's fungal research going. Still, they believed that penicillin was crucial only for studying influenza—no one, not even Fleming himself, realized that he had stumbled

Sir Alexander Fleming in his laboratory at St. Mary's Hospital in Paddington, London, October 2, 1943

upon a rare strain of fungus that produced penicillin at such an extraordinary level that it could be used to treat human infections. By the summer of 1929, just a year after its discovery, he abandoned work on the
penicillin molecule. It would be more than a decade, and another world
war, before Fleming and colleagues at Oxford would revisit it, teaming
up with the burgeoning pharmaceutical industry to create the world's
first mass-produced, commercially available antibiotic.

CHAPTER 2

A Golden Era

FLEMING AND HIS COLLABORATORS are credited with making the first widely accessible antibiotic, but that's not entirely accurate. Yes, their protocol did lead to mass production and distribution of penicillin in 1945, but it turns out that humans have been consuming antibiotics for millennia, whether they knew it or not. Significant levels of tetracycline, a broad-spectrum antibiotic still used today, have been found in the skeletal remains of Sudanese mummies dating back to AD 350 to 550. (Beer brewed at the time appears to have been the source.) And in Egypt, samples taken from femoral bones of skeletons from the late Roman period at the Dakhla Oasis also show traces of tetracycline. (It's unclear if booze was served there.) Not surprisingly, the rate of infectious diseases documented in these disparate populations has been exceedingly low. Our ancestors understood the effect of antibiotics, if not the mechanism.

Prehistoric antibiotic exposure was not just limited to Africa. Traditional Chinese medicine has provided us with an array of antibiotics, including artemisinin, a malaria drug extracted from *Artemisia* plants, which has been used by Chinese herbalists for thousands of years to treat a variety of diseases. Antibiotics are all around us; bacteria from the red soils in Jordan are still used today to treat infections. The challenge is not just finding them but also proving unequivocally that they're safe and effective for use in humans. That's where things get complicated.

Since antibiotics are omnipresent, it can be difficult to precisely define what constitutes one. Do antibiotics all have a characteristic shape? Or size? Could a drug used to treat another disease, such as cancer or gout, also serve as an antibiotic? Lots of chemicals kill bacteria—acid and bleach destroy all sorts of living things—but they're not all considered antibiotics. We're talking about substances that wipe out infections without killing us.

If we drill down a bit further, most experts use the term *antibiotic* to refer to any molecule, produced by a microbe or by a human working in a laboratory, that can be used in the treatment and prevention of bacterial infection. To qualify, it must either kill or inhibit the growth of at least one type of bacteria; those that kill are bactericidal, those that inhibit are bacteriostatic, and scientists often quibble about the distinction, because bactericidal drugs tend to be more effective. Some antibiotics can also kill parasites and fungi, but they are rarely effective against viruses. That's why your doctor is stingy about doling out antibiotics when you have a cold, as its symptoms are usually caused by a virus. (Scientists didn't appreciate that bacteria and viruses were different until 1930; viruses replicate inside of other organisms—plants, animals, humans, bacteria—and are generally impervious to antibiotics.)

The triumph of penicillin streamlined the production of dozens of new antibiotics—colistin, tetracyclines, aminoglycosides, cephalosporins—with companies racing to churn out the next lifesaving product. The 1950s would come to be known as the golden era in antibiotic development, a time when advances in molecular biology led to all sorts of new medications that markedly increased life expectancy. In fact, half of the drugs in use today were discovered during that decade.

In 1950 the Charles Pfizer pharmaceutical company's entire sales force consisted of just eight people. A year later, it was one hundred. By 1952, Americans were spending more than $100 million per year on broad-spectrum antibiotics. The economics of health care were evolving, and for the first time, leading medical journals dropped a long-standing policy against branded advertisements. In 1955, the *Journal of the American Medical Association* had more pages of advertisements than *Life* magazine, and across the country, doctors found that they could make more money working in research and development at pharmaceutical companies than from treating patients. A year later, the Viennese microbiologist Ernest Jawetz took stock of the progress. "The majority of bacterial infections can be cured simply, effectively, and cheaply," he noted. "The mortality and morbidity from bacterial diseases has fallen so low that they are no longer among the most important unsolved problems of medicine." As the 1960s approached, infections were becoming an afterthought. It was time to move on to more pressing matters like cancer and heart disease.

Some had a less sanguine view of the ways medicine was changing. In 1953, two of the leading doctors of the day, Maxwell Finland and Louis Weinstein, pointed out a few problems with all of the fabulous new medications. Antibiotics could save lives, certainly, but they could also harm almost any organ in the body, and it was difficult to predict just when that would happen. In the wrong person, they could be lethal. The duo, writing in the *New England Journal of Medicine*, urged caution, arguing that physicians should be "hesitant to employ them in cases in which indications for their use are either entirely absent or at most only slightly suggestive, lest a simple or mild disease be converted into a serious and potentially fatal one." Most notably, chloramphenicol, an antibiotic that had been used to combat potentially deadly louse-borne typhus outbreaks, became linked to aplastic anemia, or grey baby syndrome. By the end of the 1950s, the toxic side effects of antibiotics were well established, and scientists had reported a dramatic increase in antibiotic-resistant infections. Bacteria had subtly changed shape so that wonder drugs could not find them, and built enzymes to chew up any molecule that might pose a threat. The medical-industrial complex evolved during that exceptional decade; bacteria did, too.

Antibiotic research and development slowed a bit in the 1960s, as the pharmaceutical industry turned its collective attention to more lucrative diseases just as bacteria were becoming more clever. That strategic decision, along with a more rigorous drug approval pathway, prompted a steady decline in antibiotic discovery. During the downturn, prominent voices offered reassurance. In 1962, Sir Frank MacFarlane Burnet, an Australian immunologist and Nobel laureate, wrote, "One can think of the middle of the 20th century as the end of one of the most important social revolutions in history: the virtual elimination of the infectious diseases as a significant factor in social life."

A historic miscalculation had occurred. The 1970s and 1980s would become a time of tremendous advancement—scientists developed a more sophisticated understanding of everything from gene splicing to the big bang—but the years produced a notable dry spell for the pharmaceutical industry. No new classes of antibiotics were introduced during those decades.

Then a breakthrough. In the summer of 1995 a group of forty scientists published the entire genomic sequence of *Haemophilus influenzae*, a bacterium known for causing lung infections, in the journal *Science*. It was a landmark study, the first to provide us with all of the genetic information for one bacterium—the same one Fleming studied when he was first tinkering around with penicillin. The senior author on the *Haemophilus* paper was Craig Venter, a scientist, entrepreneur, and professional agitator, and the information his group produced was poised to redefine drug development. After decades of setbacks, the era of genomic medicine had finally arrived. Scientists could use Venter's information as a template to develop all sorts of drugs that had once been unimaginable.

Venter's team revealed hundreds of genes to explore as possible targets for new antibiotics, and the work generated a burst of interest from Big Pharma. GlaxoSmithKline, the London-based company, had moved away from antibiotic development, but it jumped aboard the genomic bandwagon and spent the next seven years (and close to $100 million) using genetic screens, which rely on a mix of robotics, automated detectors, and computational software to identify new drugs. Between 1995 and 2001, nearly a half million compounds were screened by GSK scientists. But only five emerged as genuine leads—and none were useful in humans.

In short, the program proved to be a massive failure. Using Venter's genetic information to screen for antibiotics had been a big waste of time and prompted a radical shift in corporate strategy. GSK began to invest much more selectively, funding teams of chemists synthesizing drug-like compounds rather than hunting for genetic targets. It began a quest to find fewer agents, and only those that could clearly justify the exorbitant costs of making them.

GlaxoSmithKline was not alone. The genomic approach was a waste for a number of companies that spent millions on similar failed hunting expeditions. The *Haemophilus* genome created an aura of excitement and investment that we now know was misguided. A decade after the first bacterial genome was sequenced, not a single drug in development was derived from that approach. The debacle set drug discovery back a generation, if not more, and the field still hasn't recovered. In the aftermath, Big Pharma has become far more conservative. Many companies simply gave up looking for antibiotics, and that has led to the troubling situation we find ourselves in today. The bacteria that cause deadly infections in humans have become quite adept at inactivating the drugs we use to treat them, and we may soon run out of options.

The US Food and Drug Administration (FDA) now approves several dozen new drugs every year, known as new molecular entities (NMEs), but very few are antibiotics. The number of patents filed for new antibacterial drugs dropped by a third (34.8 percent) from 2007 to 2012, and a 2017 survey found that only forty-one antibiotic candidates were in clinical development, compared with more than five hundred for oncology (cancer). The development of antibiotics has seemingly come to a halt just when more drugs are needed.

Dr. Anthony Fauci, director of the National Institute of Allergy and Infectious Diseases at the National Institutes of Health, is the man responsible for establishing federal funding priorities for research on antibiotic resistance, and he told me that developing new drugs is, in fact, one of his top priorities. But the situation is complicated. "You don't want the federal government to be a pharmaceutical company," he said, "because you'd have to build an entire industry, and that would divert away from what the government does well, which is scientific discovery and concept validation. We need a partner."

And that partner, for better or worse, is Big Pharma. "If the federal government tried to re-create Merck," Fauci said, "it would cost billions of dollars. The expertise of production, filling, packaging, and lot consistency. People take that for granted, but that's an art form that has been perfected by these companies, not the government."

The problem, ultimately, is that many antibiotics are not very profitable. When a new drug emerges from an idea, there's a step-by-step process that costs upward of a billion dollars to bring it to market. If that leads to Viagra, the expense is justified because you've just made a multibillion-dollar drug. With an antibiotic, however, the profit margins are narrow because of three characteristics: they're usually given in short courses, they're prescribed only when someone is sick, and sooner or later even that terrific new antibiotic is going to develop drug resistance. The latter is not a matter of *if* but *when*. "The incentive to make major investments in antibiotics," Fauci told me, "is not something that attracts the pharmaceutical industry, so how do you get around that?"

A host of options are on the table, some more plausible than others. Most involve limiting financial risk while providing financial enticements called "push" and "pull" incentives. "You can tell the pharmaceutical company, 'If you can make an antibiotic, we'll give you a tax break,'" Fauci explained, "'or we'll expand the patent on the blockbuster drug that's making you five billion dollars. We'll extend that for two years, provided you take a part of that profit and invest in developing a new antibiotic.'"

But that's not always enough. When a pharmaceutical company evaluates the overall risk/benefit and profitability of pursuing the development of a drug, it uses a metric called net present value (NPV). This is the sum of all investment costs in research and development as well as expected future revenues. For antibacterial drugs, the NPV is approximately $42 million; for most other products—including muscular and neurologic drugs—the number is closer to $1 billion. Over the past few decades, fewer and fewer companies are willing to take that risk for a relatively small profit.

Pharmaceutical research and development has the highest failure rate for new products of any industry, which raises important questions: How far should we go to incentivize the production of new drugs? Do we alter the tax code or manipulate patent law to encourage pharmaceutical companies to take a chance on developing antibiotics? Should we create an approval pathway allowing smaller clinical trials for antibiotics that treat life-threatening infections? As I pondered these thorny issues, Dr. Walsh and I got to work on our own trial—one that we hoped would tackle the expanding problem of superbugs.

Allergan has been one of the only companies still betting on antibiotics. Most of its competitors, including Pfizer, Merck, Novartis, and Johnson & Johnson, have gradually pulled funding from research and development, or simply given up. As Fauci pointed out, it's a risky process, and these companies are accountable to shareholders, not patients. Allergan is truly unusual in this respect. The company not only provides robust funding to academics like me who are developing antibiotics, they have been ramping up investment and actively scavenging for discarded drugs—molecules that other companies own but aren't using—which brings us to dalba.

In contrast with Alexander Fleming's approach, in which serendipity drove drug discovery, the team of scientists responsible for dalba didn't find the drug; they built it. It was the product of a brilliant program aimed at improving existing molecules by using molecular modeling and simulation exercises to create better drugs. This, in short, is rational drug design: chemists use computer simulations to model how existing chemicals might interact with dangerous agents of disease, or pathogens. Add a carbon atom here, remove a nitrogen atom there, and see what happens. This field, often populated by the most innovative programmers and scientists, is called synthetic organic chemistry, and it's wildly expensive.

Antibiotics that have been around for years can be rearranged to make them even stronger. Molecules are ripped open and split apart, then they're fermented, ionized, reassembled, and purified. It's a bit like a chef gunning for an additional Michelin star: chemists tinker with the microscopic ingredients until they find the perfect recipe. I like to think of the test tube as a tiny sauté pan.

Dalba was made by extracting a large molecule, known cryptically as A40926, from a bacterium that was found in Indian soil in the 1980s during a hunt for antibiotics. Chemists carefully removed clunky appendages from A40926, including sugars, amino acids, and carbon atoms, but they were careful not to tamper with a small cleft in the molecule, known as a binding pocket. That pocket would allow their sleek new drug to seek out and destroy the cell walls of bacteria while avoiding human cells. This property is what would turn A40926 into a trimmed-down, powerful antibiotic called dalba. They asked Tom Walsh and me to test their recipe.

Chemical structure of dalbavancin

I have often wondered why Allergan was doubling down on antibiotics, but I wasn't naive enough to think the investment was altruistic. The company has successfully ushered a number of drugs to market, including a potent antibiotic called ceftaroline, but it's not widely used because it's so expensive; the cheaper generic alternative works just fine. My hospital actually refuses to carry some of its drugs because of the price. Still, Allergan turns a healthy profit because it owns Botox, for which there is no alternative; the drug did around $3 billion in sales in 2018.

The rise of superbugs irrevocably changes the calculus. Patients and doctors are rightfully terrified, and if Allergan can develop a new treatment that works—something that saves people with potentially lethal infections—the company knows we'll have no choice but to use it. As a society, we'll pay just about any price.

I opened the thick Manila folder Tom Walsh had handed me. The first few pages were internal documents of Allergan boilerplate, but page 7 caught my eye. It was a brief history of dalba. In December 2001 Lehman Brothers had predicted a 2005 launch for the drug, with sales of $7.7 million that year. From there, the drug was expected to go viral, generating $65.8 million in 2006, and $225 million in 2008. But, like so many things emanating from the investment bank, the prediction was wildly inaccurate. Pfizer spent years trying to show that dalba could be used safely to treat superbugs, but in 2007 the FDA issued a statement saying it wasn't convinced. More data were needed before the antibiotic could be approved. In 2008 Pfizer threw in the towel, withdrawing all marketing applications for the once-promising drug. Durata Therapeutics acquired it one year later, and that Chicago-based company was eventually purchased by Allergan—along with the exclusive rights to dalba. After extensive follow-up, the FDA finally approved dalba to treat skin infections in May 2014, but doctors weren't comfortable using it.

An antibiotic that saved lives in 2011 might not work just a few years later, and anxious clinicians didn't want to take a chance when lives were on the line. Superbugs were evolving in ways we never expected, creating thousands of enzymes to chop up and destroy antibiotics. They were also developing molecular machinery known as efflux pumps to excrete antibiotics, rendering the drugs useless. With a single mutation, bacteria can spoil the chemists' recipe, and the delicately designed antibiotic is ruined. Dying patients could be given something that simply doesn't work, and a billion-dollar investment evaporates.

These genetic mutations are difficult to detect; doctors and patients usually don't know about them until an infection has taken hold or started to spread. Sometimes we don't discover the mutation until the autopsy. I had attended dozens of lectures and workshops on antibiotic development, but no one mentioned that bacteria were mutating so fast that even the most spectacular new products couldn't keep up. It was the best-kept secret in medicine.

In order to appreciate how we arrived at this crossroads—and to understand why a mechanic from Queens named Jackson might die from his infection—it's useful to know how antibiotics were first used and, more importantly, misused. The history of human experimentation is unnerving and unpleasant, but it informs how clinical research is performed today, and it explains why my dalba trial was going to be so difficult to pull off.

PART 2

First Principles

CHAPTER 3

The Lucky Grenadier

ON OCTOBER 24, 1914, as Alexander Fleming was pouring antiseptic fluid into the bloody leg of an injured man in Boulogne, a German soldier on the opposing side of the war was busy composing a letter. "Dear parents," Gerhard Domagk wrote from a coastal town in Belgium, "I have just arrived in Oostende with a lot of other platoons of volunteers. Soon after we had left Bruegge behind, we heard the rolling of thunder of our naval artillery. Apparently we are about 12 km away from the battlefield."

Domagk, a tall, thin, anxious young man with a high forehead and light, penetrating eyes, was far closer to danger than he realized. Like so many of his medical school classmates, Domagk had been swept up in a wave of nationalism after Germany and Russia declared war on each other just a few months earlier, and had put his studies on hold to join the effort. In the span of ten weeks, the nineteen-year-old had been inducted into the Leibgrenadier Regiment of Frankfurt on the Oder, a unit specializing in grenades, given brief training, and shipped off to fight. He was now stationed one hundred kilometers from Fleming's makeshift hospital and perilously close to the Western Front.

As he put pen to paper on that chilly October day, medical school was a distant memory. Domagk had traded his white coat for weaponry and was now camped out near the same muddy combat zone where so many soldiers were contracting tetanus. In a few days, his regiment would be ordered to participate in a simple yet terrifying tactical maneuver: to charge forward with grenades and drive the Allies from their trenches, which were located just a few miles away.

It was a suicide mission. Most of Domagk's unit was killed within the first few minutes of the surprise attack, gunned down in a hail of bullets as they tossed their grenades into the air. The Germans would lose over 135,000 soldiers on that battlefield, many of them students. Historians call it the First Battle of Ypres, but Germans remember it as *Kindermord*: the Massacre of the Innocents.

Gerhard Domagk survived. He hastily retreated with his comrades to the Eastern Front, where they dug in and waited for new marching orders. Several weeks later, just before Christmas 1914, his helmet was knocked off his head by enemy fire, rendering him unconscious. But once again, he survived. Given his abridged medical training and heavy German casualties, Domagk was reassigned to care for injured soldiers. The field hospital was a farm in the middle of the woods in Ukraine, where tents served as hospital wards, a barn had been converted into an operating room, and farm carts functioned as ambulances. He would serve there for two years.

The young grenadier was now a triage medic, responsible for sorting the arrivals, quarantining cases of cholera, and identifying the less severe wounds that might benefit from surgery in the barn. He also transported the dead and dying to mass graves. There, in that dense forest, Domagk observed the same festering shrapnel wounds that Fleming was seeing in Boulogne, and the experience would leave a similarly profound impression.

After the war, Gerhard Domagk returned to medical school at Kiel University, in northern Germany. Like Fleming, Domagk was a talented medical student with a passion for performance art (Domagk loved music, Fleming was devoted to drama), and both men, pacifists by nature, were haunted by visions of death and destruction. But the theater of war would propel both men to greatness.

In 1923, two years after graduating from medical school, Domagk attended a meeting that changed his life. At a gathering of the German Society of Pathology, he met his future mentor, Walter Gross, a scientist who helped him realize that, like Fleming with Almroth Wright, he was more comfortable with microscopes and microbes than with treating patients. Domagk was a principled man, a high-minded thinker with firm beliefs about right and wrong, good and evil, and he wanted to understand the underpinnings of life. This, he believed, would bring him closer to humanity, and it drove him toward research.

The Great War had left him shaken, and in an effort to move on from the trauma, he created a guiding principle: "Whatever contributes to the preservation of life is good; all that destroys life is evil." He accepted an invitation to join Gross's research outfit and began studying how blood and liver cells helped the human body fight infection.

The lucky soldier proved unlucky in the laboratory. Although Domagk was known for his keen eye—he could spot four-leaf clovers in a field—he was unable to strike upon a big scientific discovery, one that would secure his reputation and, more importantly, permanent funding. He had trouble supporting his family as a middling academic. (The family motto during that time was "Work, work, and go hungry.") He eventually took a job with IG Farben, the largest chemical plant in the world.

In those days, German academia had close ties with the nascent pharmaceutical industry, and, in contrast with other industrialized nations, universities often catered to the needs of the private sector. Curricula were tailored to workforce demands. Unlike other industries, pharmaceutical companies had minimal overhead, requiring little in the way of equipment and lab space. What really mattered was human capital: scientists with the expertise to discover and develop drugs.

The sprawling company was situated near the eastern bank of the Rhine and employed nearly two thousand workers in hundreds of buildings. It had its own bank, library, printing press, and fire department. The architecture of the ersatz town was decidedly modern, a glistening mash-up of uncluttered concrete façades and undisturbed parkland on a small hill rising just above the river. Domagk's boss, Carl Duisberg, was a confident man with a thick, brown mustache and wire-rimmed glasses. He was a forceful CEO, aggressively pursuing mergers, acquisitions, and intellectual property to suit the needs of his ever-expanding company. What started out as a small dye-making plant had grown, under Duisberg's stewardship, into a pharmaceutical behemoth that became the largest corporation in Germany.

In 1927, Gerhard Domagk, age thirty-two, signed a two-year renewable contract with IG Farben to search for undiscovered drugs. It didn't so much matter what kind of product he looked for—Domagk was free to look for molecules to treat gangrene, diarrhea, or pneumonia—as long as the end result could turn a profit. The hunt was via chemical screens, where thousands of similar molecules were tested against bacteria in a variety of animals, including mice, rabbits, and occasionally canaries. If a new drug killed bacteria but not the animal, it was considered a success, and led to more questions and further testing. Could it be used in humans? At what dose? What were the side effects? It was tedious work, but with Duisberg's vast resources it had tremendous potential.

Domagk spent most of his days inoculating mice with *Streptococcus*, a bacterium that could cause pneumonia, meningitis, endocarditis— and gangrenous battlefield wounds. The mice were then given new chemicals orally, intravenously, or as an injection. Thousands of mice died at Domagk's hands, and he performed autopsies on the carcasses, searching for signs that the strep infection might have been partially defeated. But there were no hits.

For five years, Domagk tested thousands of chemicals. Then someone in his lab had an idea. Chemists at IG Farben had made wool dyes resistant to fading by adding sulfur atoms. If sulfur-containing molecules stuck more tightly to wool, might they also adhere more tightly to bacteria? It was an odd suggestion, a result of the outside-the-box thinking that was encouraged at the company, and it was something that smaller competitors could not do. The project was greenlit.

In October 1932 a member of Domagk's team added a sulfur-containing molecule called sulfanilamide to an existing dye to create a compound called Kl-695. Sulfanilamide had been around for years as a cheap and practical dye, but it had never been tested for antibacterial properties.

Chemical structure of sulfanilamide

In the fall of 1932, Domagk's assistant began giving Kl-695 to animals. The results were striking: the chemical protected mice from *Streptococcus* and produced no side effects. Additional experiments yielded even more intriguing results: attaching sulfanilamide to *any* dye transformed it into an antistrep drug. Chemists at IG Farben continued to tinker with Kl-695, adding and subtracting atoms to alter the molecule's activity. One variant, Kl-730, was stronger than anything they'd seen before. Mice that received the drug became impervious to *Streptococcus* at almost any dose. Domagk was asked to present his findings to his marketing team and patent attorneys, and Kl-730 was rebranded as Streptozon. After striking out for years, Domagk finally had his hit.

But there was a problem: the patent for sulfanilamide had expired. Once Domagk's experiments were made public, anyone with a working knowledge of chemistry could make this new wonder drug. He knew that publishing his work would be like tossing a grenade, destroying the antiseptic school approach to treating infections—while securing his reputation as a brilliant scientist—but IG Farben couldn't patent every variant. Domagk and his team hastily tested as many as they could and filed a patent for the most potent one, Kl-730, on December 25, 1932, thirty-six days before Adolf Hitler would become chancellor of Germany.

———————

IN OCTOBER 1939, one month after Nazi Germany invaded neighboring Poland, igniting World War II, Hitler issued his order involving *Gnadentod*, or "charitable death," and weeks later gas was used for the first time to kill innocent people in Poznań, Poland. Hitler had made it known to doctors, including Gerhard Domagk, that it was necessary to get rid of "useless mouths"—the young, the old, the disabled, the terminally ill—and most complied, loyal servants to Nazi Germany. In fact, the professional group with the largest percentage of Nazi Party members was medicine. It wasn't until after World War II that the public would discover exactly what that meant and the extent to which doctors had experimented with Domagk's chemicals to facilitate Hitler's mission.

The revelations began during the Nuremberg war crimes trials, which were held between November 1945 and April 1949. Nuremberg held special significance for the defeated Germans: the city had been the location of Nazi Party rallies, and it was where laws stripping Jews of their citizenship had been passed. There was symbolic value in making Nuremberg the place where Nazis would face justice.

The International Military Tribunal included judges from the Big Four victorious powers: the United States, France, the United Kingdom, and the Soviet Union. Twelve subsequent proceedings were held before military tribunals and American civilian judges. It was during this time that the world would appreciate the grotesque role that antibiotic experimentation played in the Final Solution. Of the twenty-three defendants from case no. 1 of the subsequent proceedings, eight were members of the medical service of the German air force, seven were members of the SS medical corps, and eight were prominent health care providers within the Nazi Party.

In the courtroom, most wore German army uniforms stripped of insignia. In their opening remarks, lawyers for the prosecution explained how moral disintegration had transformed these distinguished doctors and scientists into assassins. The defendants had performed their work under the guise of medical research, but internal documents

*Nazi war criminals in the dock during
the Nuremberg trials*

revealed that they knew that their experiments were not scientific and did not follow established protocol for human studies.

At the arraignment, US brigadier general Telford Taylor, chief of counsel for war crimes, read the charges aloud: the accused had performed experiments at concentration camps that involved freezing, burning, drowning, and poisoning. They had also conducted sterilization and transplantation experiments on the subjects they called *Versuchspersonen,* or "experimental persons." The doomed had also been gashed and gouged, then infected with bacteria or ground glass or sewage, before receiving varying doses of Gerhard Domagk's prized molecule, sulfanilamide.

By the start of World War II, Domagk's team had been combining sulfanilamide with other molecules, which were sold under various names

*Brigadier General Telford Taylor,
Chief Counsel for the prosecution,
is shown reading the charges against
the twenty-three Nazi doctors at
the opening of their trial before the
Allied Military Tribunal
in Nuremberg, 1946*

to treat infections. (Prontosil was the most popular.) But they weren't without their dangers: patients routinely developed nausea, vomiting, rash, and kidney failure while taking these sulfanilamide derivatives. Some had even died.

The Nazis needed something to treat their injured soldiers. They had suffered heavy casualties from bacterial infections on the front lines during the winter of 1941, many due to *Streptococcus*. Germans had heard that the Allies were using a miracle drug called sulfanilamide—a drug Germans had discovered!—to protect their fighters from gangrene and sepsis and pressed their medical officers to do the same. But Nazi physicians weren't convinced that sulfanilamide was truly safe or effective.

The Nazi Party's early relationship with IG Farben was chilly. Nazis had condemned animal research, which was integral to the company's drug development strategy, as a form of "Jewish science" that was linked to ritual slaughter, and some of the sulfanilamide researchers did not support the Third Reich. Gerhard Domagk was not a Nazi—he refused to return the Heil Hitler salute—and the führer forbade him from accepting the Nobel Prize in Medicine when it was awarded to his team in 1939. After the prize was announced, the Gestapo arrested Domagk and threw him in jail for a week. His values never wavered, however. The scientist clung tightly to his guiding principle. But by 1941, many top officials at his company were more than willing to assist with the experiments.

Members of the Military Tribunal learned in court that sulfanilamide experiments began at Ravensbrück women's concentration camp in northern Germany on July 20, 1942. Sixty women, divided into five groups of twelve, and fifteen men were experimented on to determine how to bring about battlefield infections such as gangrene and whether sulfanilamide could prevent or treat them. Four-inch-long incisions were made into the leg muscles of these prisoners. Wood shavings were combined with sugar and bacteria, and the mixture forced into the open muscle. The wound was then closed and a cast was applied. In a second group, the procedure was repeated, but wood shavings were replaced by fragments of glass. In a third group, both wood and glass were used. The prisoners were then given varying doses of sulfanilamide.

Because none of their subjects died, Nazi physicians increased the severity of the infections by cutting off circulation to the wound, which prevents the immune system from fighting bacteria. This time many deaths followed. The few women who survived were called *Kaninchen*, German for "rabbits," because they hopped or limped around Ravensbrück. In the courtroom at Nuremberg, four Polish survivors testified that they would have preferred death to the defleshing experiments. A

Gerhard Domagk

physician from Boston, Leo Alexander, testified that sulfanilamide tests could have easily been performed on soldiers with actual battlefield infections. Nazi physicians carried out their barbarous experiments simply because they could. The twenty-three defendants all pleaded not guilty. Sixteen were found guilty, and seven of those were sentenced to death by hanging.

Toward the end of the trial, a medical professor named Andrew Ivy was summoned to testify. Dr. Ivy was a well-known physiologist and ethicist and had been nominated by the American Medical Association (AMA) to the US secretary of war to be an expert witness for the prosecution. He was founder of the Naval Medical Research Institute, where he studied altitude sickness and tried to create potable seawater. Ivy's testimony was steadfast—under cross-examination, he said there was "no politician under the sun that could force me to perform a medical experiment which I thought was morally unjustified"—and laid the foundation for what would become the Nuremberg Code, a ten-point framework for conducting future experiments on human beings.

On the witness stand, Professor Ivy read from a set of principles and rules set forth by the AMA for clinical trials. These included: (1) that the individual upon whom the experiment was to be performed must give voluntary consent, (2) that the danger of the experiment must be previously investigated by animal experimentation, and (3) that the experiment must have appropriate supervision.

"To your knowledge," the prosecutor asked Dr. Ivy, "have any experiments been conducted in the United States wherein these requirements which you set forth were not met?"

"Not to my knowledge," Dr. Ivy replied.

He could not have been more wrong.

CHAPTER 4

Embedded

SEVERAL WEEKS BEFORE Gerhard Domagk discovered the promise of sulfanilamide, the US government gave an American physician named R. A. Vonderlehr an unusual assignment. In the fall of 1932 the thirty-five-year-old was tapped to move to Tuskegee, Alabama, to assemble a group of adults for an experiment. Vonderlehr was tasked with recruiting black men between the ages of twenty-five and sixty who had untreated syphilis, an infection caused by a corkscrew-shaped organism that can reveal itself as anything from a painless rash to life-threatening meningitis. These Alabama sharecroppers would serve as human guinea pigs, prodded and probed with blood tests, X-rays, and, finally, a spinal tap to determine if syphilis had reached the brain. Dr. Vonderlehr's experiment was to last six months, and, unbeknownst to the infected men, he had no intention of providing treatment, making it both a cheap and simple project.

At the time of his assignment, the treatment for syphilis consisted of a cocktail of arsenic and mercury; the cause and cure were once described as "a night with Venus and a lifetime with Mercury." To prove even mildly effective, the preparation had to be given several dozen times over the course of a year, and relapses were common. Vonderlehr's project, an extension of ongoing work in the area, set out to discover what happened to people with syphilis who did not receive treatment. This clinical investigation, he believed, would be an important step forward for the study of infectious diseases and for government-funded human experimentation.

An expert in dermatology and syphilology, Vonderlehr was among the most gifted officers in the US Public Health Service Commissioned Corps, and one of the few in the country with subspecialty training in cardiovascular syphilis, a potentially catastrophic complication of untreated infection. He was an ideal candidate to serve as the on-site director for the Tuskegee Study.

When he headed south, approximately thirty thousand people lived in Macon County, Alabama, where Tuskegee was located. Eighty-two percent were black. Most of the houses in the area, known as the black belt for its rich, dark soil, lacked proper floors or indoor plumbing. Twenty-two percent of black residents could not read or write. (By contrast, the illiteracy rate was just 2 percent for whites.) Tuberculosis, malnutrition, and vitamin deficiencies were common, and more than one in three adult men had syphilis. Few of the black belt inhabitants received routine medical care, which meant that a smoldering infection such as syphilis often went untreated.

Vonderlehr arrived in Alabama in the fall of 1932, around the time that Domagk and his team discovered that sulfanilamide protected mice from *Streptococcus*. His first objective was recruitment. To attract patients, he traveled to black schools and churches to announce that government doctors had arrived to offer free blood tests. Eager for medical attention, the impoverished residents of Macon County turned out in droves to participate in Vonderlehr's study, which was performed in conjunction with local officials at the Alabama Department of Health, the Macon County Department of Health, and the Tuskegee Institute (now Tuskegee University), known at the time as the Negro Harvard.

As part of the state and federal funding arrangement, a local black physician and a black nurse assisted in patient recruitment and the initial evaluations. Embedding locals was a novel form of collaboration; it was thought to serve as a potential road map for future studies involving human research subjects. The partnership helped government officials acquire something that was essential for any clinical trial: trust.

Qualifying subjects met Dr. Vonderlehr, who performed a physical examination and obtained blood samples that were shipped to the state health laboratory in Montgomery. By the end of the first week, three hundred samples had been collected. Men who tested positive were notified by mail and instructed to return for further testing. But patients were not informed that the tests were for syphilis; rather, they were told they had "bad blood," which locals used to describe a variety of ailments, from headaches to indigestion.

There was no shame in participating because researchers had not explained to the sharecroppers that *bad blood* actually referred to a contagious, potentially fatal disease that is almost always sexually transmitted and can cause birth defects. (Women were excluded from the experiment because it was assumed that they would not be forthcoming about their sexual histories.) Deception led to confusion: sick-appearing men were routinely told they did not have bad blood, while healthy-looking men were told they did. Vonderlehr, afraid that patients would drop out if they understood what a spinal tap entailed, rounded up the sharecroppers in large groups to get the procedure done quickly, before word could spread of the substantial risks. "The details of the puncture," he told medical colleagues, "should also be kept from them as far as possible."

As word spread of the pain and debilitation associated with the spinal tap, Vonderlehr enticed the men to undergo the test by suggesting that it might serve a therapeutic purpose. "This examination is a very special one," he wrote to his patients in a form letter, "and after it is finished, you will be given a special treatment if it is believed you are in a condition to stand it." An associate complimented Vonderlehr on his "flair for framing letters to Negroes."

The ruse worked, and soon the doctor was performing up to twenty spinal taps a day. Toward the end of the exhausting six-month study, Vonderlehr found himself enthralled by reams of data and requested an extension. Five or ten more years of observation, he said, would provide invaluable insights into the natural course of untreated syphilis. But his supervisors were not convinced and recommended terminating the study as planned in May 1933.

Then fate intervened. A month after the scheduled termination, Vonderlehr's boss retired and the doctor was promoted to director of the Division of Venereal Diseases. With Vonderlehr at the helm, the study would go on, and, more importantly, there would be no end date—not until the last enrollee had died. That, he knew, would take decades. With that decision, the Tuskegee sharecroppers transformed from patients into *Versuchspersonen*—experimental persons—frozen in time, isolated from the remarkable medical advances that were about to occur.

VONDERLEHR'S STUDY continued through the 1930s and the Great Depression, but World War II presented a problem: many of the Tuskegee men tested positive for syphilis while enrolling in the armed forces, and treatment was mandated before joining the war effort. Dr. Vonderlehr stepped in, contacting the chairman of the local draft board to ensure that his subjects remained exempt from proper medical care. The men were shipped off to battle.

In the middle of the war, a major therapeutic advancement occurred: doctors began treating syphilis with Fleming's miracle drug, penicillin. This development, government officials argued, made the continuation of the Tuskegee Study of Untreated Syphilis in the Negro Male even more important, because it could never be repeated. Andrew Ivy and other influential physician-ethicists were pushing for more formal guidelines regarding human experimentation that required not just consent but *informed* consent. Now that an effective treatment was available, the sharecroppers deserved to know about it. But that would compromise the study. Penicillin would cure them, derailing the trial.

Largely shielded from oversight, the decision to proceed was an easy one. The untreated Tuskegee men would be observed and monitored until death. Unaware of the decisions being made about their care, the sharecroppers were loyal participants; during the first two decades of the study, 144 of 145 Tuskegee families consented to autopsies. And on it went like this for decades.

The Tuskegee men were not denied treatment. It was simply never offered. Despite the fact that untreated syphilis shortened life expectancy by 20 percent, a committee at the CDC (back when the acronym still stood for Communicable Disease Center rather than Centers for Disease Control and Prevention) concluded in 1969 that the experiment should continue. Roughly one hundred men died as a direct result of complications from untreated syphilis, and the participants were never told of the purpose of the study.

Forty years after Ray Vonderlehr had been given his assignment, the Tuskegee experiment finally fell apart. When the story broke in July 1972, after a former Public Health Service employee leaked the details to the Associated Press, outrage was swift. Not only had untreated sharecroppers died young, but also they had spread syphilis to their wives and children. Government researchers, fully aware of the risks, had never bothered to obtain informed consent. A lawyer representing the men called the experiment "a program of controlled genocide."

At first, the study was described in newspapers as clandestine. But this was not true. What happened in Tuskegee had been openly discussed at medical conferences and written about by physicians for decades. The first paper, in 1936, was read at that year's American Medical Association annual meeting and revealed that untreated syphilis greatly increased the risk of heart disease. In 1955, doctors reported that roughly a third of the sharecroppers died directly from syphilitic lesions in the heart or brain, but the findings did nothing to change the direction of the work in Tuskegee. In return for burial insurance and warm meals on examination days, hundreds of poor, illiterate men had agreed to participate in a government-sponsored study that irreparably harmed them and their families, and the world was once again forced to confront an uncomfortable reality: doctors don't always act in their patients' best interests.

As the scandal unfolded, Senator Edward Kennedy of Massachusetts called congressional hearings, an *ad hoc* advisory panel was assembled to examine the study, and blame was scattered across a host of institutions, including the Alabama Department of Health, the Macon County Department of Health, the Tuskegee Medical Society, and, of course, the Public Health Service. Why hadn't the study been stopped, or modified to incorporate penicillin? Why not tell the sharecroppers what was happening to them?

For the first time, Americans took an interest in the nuances of clinical research and the safeguards surrounding human experimentation and demanded answers. This wasn't simply the work of a rogue researcher; it was state-sponsored abuse. As outrage grew, government officials were compelled to produce a formal research protocol to explain how this ghastly experiment had been drawn up and allowed to continue. But here was yet another surprise: there *was* no protocol.

CHAPTER 5

Safeguards

"IT'S ALL ABOUT THE PROTOCOL," Tom said to me. We were in his office, about to discuss our dalba trial, and it was impossible to ignore the intense scrutiny we were under. Introducing a drug into a hospital to treat a deadly infection was no small task, and we were required to spell out every aspect of the experiment before it could even be considered—and there was no guarantee it would be approved. Tom was in his element here; he was drawn to challenges, lunging toward difficult cases and complicated trials. He pointed to a blank legal pad and started scribbling. "I want you to take the lead here," he said, looking up at me. "You should write the protocol."

Dalba is unusual in that you don't take it by mouth or intravenously over several hours. It's given as a onetime infusion over thirty minutes, and it's designed to have an exceedingly long half-life, so it stays in your body for weeks. This feature gives it a unique chemical profile—the liver doesn't metabolize it, and the kidneys don't rapidly excrete it—which is one of the reasons we were betting on its effectiveness. But what if we were wrong?

Staring at Tom's legal pad, a number of uncomfortable thoughts swirled through my head: What if superbugs had already mutated to withstand dalba? What if the FDA had been right to initially reject it?

We might expose patients to something ineffective. Or, worse yet, harmful. Rigorous safeguards are in place for a reason, but they can fail. "I'm ready," I said, running my hands through my hair. "I think."

Historically, physicians have governed themselves through peer review, but the horrors of Tuskegee laid bare the limitations of this approach: doctors consistently signed off on a trial that caused birth defects and deaths. In 1970, Dr. James B. Lucas, assistant chief of the CDC's Venereal Disease Branch, wrote that the Tuskegee Study would not "prevent, find, or cure a single case of infectious syphilis or bring us closer to our base mission of controlling venereal disease in the United States"— yet he concluded that the trial should continue.

Even well-meaning physicians have blind spots, and those lacunae can place patients at grave risk. To properly draft a protocol for our dalba trial, I would need to choose my words carefully, and I would need to draw on the lessons learned from prior generations of clinical researchers to understand how easily things can go wrong. But so many mistakes had been made that it was hard to know just where to start. After casting about for a few hours and bouncing dozens of ideas off of Tom, I began my assignment by traveling to the hospital's medical library. There I would write my protocol.

Eventually I figured out where to begin: a short manuscript published a half century ago. In 1966, six years before the Tuskegee story broke, a Harvard anesthesiologist named Henry Beecher published a paper in the *New England Journal of Medicine* outlining twenty-two American studies in which patients had served as experimental subjects without informed consent. It was a hot topic: consent had entered the public's consciousness after social psychologist Stanley Milgram conducted his "obedience to authority" experiments at Yale University in 1961, famously tricking research participants into believing they'd administered electrical shocks to strangers simply because a teacher had instructed them to do so. Beecher was troubled by the state of human experimentation in the United States and argued for greater oversight by pointing out how clinical research had gone awry. It was a necessary starting point.

In many of the cases Beecher identified, participants simply did not know they *were* subjects. For some, antibiotics such as penicillin and sulfonamides were withheld; in others, live cancer cells were injected into patients who were otherwise cancer-free. Perhaps the most appalling example involved the administration of a virus known to cause hepatitis into "mentally defective children." Many participants died in these unethical, outrageous studies, which took place at leading medical schools, governmental military departments, private hospitals, the National Institutes of Health, and Veterans Administration hospitals all over the country. Beecher's work underscored an uncomfortable fact: Nazis were not the only ones who produced physicians capable of abuse.

Henry Beecher knew firsthand just how easy it was to manipulate patients. During World War II, he had served in military field hospitals in North Africa and Italy, where pain medications such as morphine could be hard to find. He noticed that nurses were able to calm injured soldiers with injections of saline when they were administered as if they were shots of morphine. A single infusion of salt water enabled young men to tolerate agonizing surgeries without anesthesia, and it introduced the young doctor to the power of the placebo effect, in which a patient's belief in a treatment could be just as powerful as the drug itself.

When he returned to Harvard after the war, Beecher continued to study the placebo phenomenon and argued for a new model of clinical research: one that called for randomization so that study subjects wouldn't know if they were receiving a real treatment (like morphine) or a sham (like saline). His insights ultimately contributed to our appreciation for the randomized, controlled trial—the standard in human experimentation today—but he is best known for his eponymous report.

"An experiment is ethical or not," Beecher wrote in his conclusion, "it does not become ethical *post hoc*—ends do not justify means." He called on the "intelligent, informed, conscientious, compassionate, responsible investigator" to protect human subjects, but clearly that wasn't enough. Many of the researchers he singled out believed their experiments were ethically sound. I made a point to hunt for these blind spots in drafting my own protocol.

Despite the fallout from the Nuremburg trials, patient abuse and exploitation were still so pervasive in the 1960s that something drastic had to be done. The Beecher Report was a seminal moment in clinical research: it was the focal point around which the rules of human experimentation began to coalesce.

The same year that the Beecher Report became public, US surgeon general William H. Stewart announced that academic medical centers had to set up human-subject review committees if they wanted to continue to receive federal funding. This policy gave birth to the modern institutional review board (IRB), which fundamentally altered the way research would be conducted in the United States. Physicians were no longer free to govern themselves; instead, they had to bring experiments and protocols into the light of day, where independent experts could evaluate, revise, and potentially reject their proposals. The

Dr. Henry K. Beecher

IRB created a mechanism to protect patients, especially those who might easily be exploited. Researchers often turn to the marginalized and powerless as study subjects, and the IRB was designed to give a voice to the voiceless.

This novel expectation of internal review marks the beginning of modern bioethics—a time when philosophers, theologians, lawyers, scientists, and laypeople came together to help determine right from wrong in clinical research. One of the early successes was passage of the National Research Act of 1974, which required that nearly all research on people be vetted by an IRB before it could begin. My protocol would be submitted to an IRB, and it had to be written in a way that the board members could understand.

IRBs vary by hospital—there are currently about three thousand in the United States—and they're composed of a diverse group of individuals to represent the broad interests and values of our society. Like a sequestered jury, they deliberate behind closed doors, under a shroud of secrecy. IRBs are, in theory, an invaluable mechanism to protect vulnerable patients, but they can be a thorn in the side of impatient researchers like me. They decide who may be studied and how. They can vastly alter clinical trials or shut them down completely.

CHAPTER 6

Variables

TOM AND I regrouped a few days later. As I prepared to show him an outline of the dalba protocol, he took a moment to put things in perspective. "I know you're excited," he said, pointing at my new pens and notebook. Sitting across from him, I felt the way I had before baseball games, trying my best to fend off self-doubt while sizing up a new opponent. Excited, yes, but also nervous. I nodded, but I knew I wasn't convincing. Tom put his hand on my shoulder. "Matt," he said as his face dropped and his demeanor changed, "we bring two things to the bedside: compassion and science. If we don't get this right, they'll shut this trial down before we even get started."

I had spent long nights reviewing the science behind dalba and was convinced it would work. I didn't want to slow down. I pushed the outline across the table. "What do you think?"

He picked up the paper and wandered over to his computer and pulled up some music. Soon, Vivaldi's *Four Seasons* filled the room. "Let's take a look," he said. I was eager for his approval, and he knew it. After a few minutes, he stood up and walked over to his bookshelf. "I just thought of something," he said. "This may help." He selected a book and brought it back with him to his seat.

Tom's off-the-cuff thoughts routinely became medical doctrine years later, after everyone else had caught up, but that delay was no longer acceptable. Every week, we saw patients who were dying of infections that had once been treatable, and the protocol was my chance to convey the gravity of our work. "Does it all make sense?" I asked.

IRBs convene without researchers or study subjects, so I would not be present to explain my rationale or answer questions in person. Some reviewers might not even know what a superbug is, and I doubted that any of them had heard of dalba. Still, I had to explain the urgency of the trial in a way that wouldn't appear hurried or careless. I needed to capture what it was like to be in the room with that terrified mechanic from Queens. "I tried to make it as straightforward as possible."

Tom again tapped his legal pad as he read. My first objective was to determine a primary endpoint, or variable, for the study. After distilling the information about superbugs and dalba and drug development, I needed to address a deceptively simple question: *What is our study measuring?* "Length of stay," I offered. "Kind of an odd choice," I conceded, "but go with me on this."

In 1983, Medicare implemented an inpatient payment system that altered the way medical centers do business. Rather than simply reimbursing whatever costs they charge to treat patients, hospitals were going to be paid a predetermined lump sum per patient based on something called the diagnosis-related group (DRG). Simple diagnoses such as pneumonia might generate a few thousand dollars, whereas an organ transplant could be a few hundred thousand. Hospitals were quick to respond to this radical restructuring of financial incentives: the average length of stay for a patient dropped from 10.0 days in 1983 to just 5.1 days in 2013. Turnover increased, and so did revenue.

The faster patients are discharged—the shorter their lengths of stay— the faster the hospital can admit another patient and take in another lump sum. The length-of-stay metric is so important that I receive a printout every quarter informing me of how long my patients stay compared with those of my colleagues. It is one of the only head-to-head comparisons that I ever receive.

"We're going to show that patients are cured," I said to Tom, "*and* that they have a shorter length of stay." He put down the outline, and I made my case: length of stay would be the focal point; it would drive every decision I would make about the trial. Tom nodded and handed the outline back to me. I returned to my office, imagining the hypothetical group that would soon judge our protocol.

IRBs must include five people, but they can be any five people, provided that one is not affiliated with the institution and another represents "nonscientific" concerns—often a member of the clergy. Expertise is not a prerequisite. Their job is to make sure that the people being studied understand the risks and benefits of participating and aren't selected in discriminatory ways. Putting together the protocol would be a new experience for me. I had templates to work from, but I wanted the document to be something original.

Over the next few weeks, I continued my daily responsibilities at the hospital, caring for patients with superbugs and other maladies. Nights were spent writing the protocol. I went through dozens of drafts, trying to explain what I was seeing in the hospital, trying to convey the gravity of the moment, looking for inspiration amidst the desperation. Once we agreed on the wording, I would submit it to the IRB for approval.

Several weeks later, after a string of meetings that went late into the night, I had completed our protocol: fifty-eight pages and more than fifteen thousand words spelling out what we hoped to do. We were going to see if dalba was more effective than existing treatments for patients admitted to the hospital with a skin or soft tissue infection attributable to a superbug or some other newly mutated bacteria. The trial was going to be split into two parts: a *pre*period and a *post*period. For the first six months—the preperiod—I would observe how patients with serious skin infections responded to standard medical care. Most would be battling superbugs, and all would require hospitalization. This would be followed by a six-month postperiod, where new patients with the same infections would receive dalba instead. NewYork-Presbyterian employs more than six thousand physicians who care for more than two million patients every year, and we would be the first doctors to use dalba. The whole thing was supposed to take about a year—an annus mirabilis.

Tom and I had one more issue to resolve. The trial needed a principal investigator, known as a PI, to serve as the lead researcher. This person oversees every detail of the study and is ultimately responsible for its success or failure. Tom suggested that I should run the trial and he could serve as my lieutenant. With a handshake, we formalized the deal, and I submitted our protocol to the IRB. From here on, I was in charge.

And that's when the trouble began.

Chapter 7

Deferment

I MARCHED INTO Tom's office with a handful of papers. *This is ridiculous,* I said to myself. I grabbed a chair and dropped my papers onto his desk. The IRB had refused to approve the clinical trial.

"Deferred," I added. "What does that even mean? The trial was deferred? Til when?" The IRB had sent me a four-page document pointing out problems large and small with the protocol. I tried to stifle my frustration, but it was impossible. I hadn't just been deferred, I'd been rejected. I couldn't get through the letter without thinking about Henry Beecher, my inspiration for the protocol, and how he exposed the nefarious investigators of prior generations. I was just like them: a physician with blind spots—or worse. Beneath my anger, I was ashamed to admit, was embarrassment. Tom had given me an opportunity—writing the protocol was something he could easily do himself—and I had fumbled it. I held up the document and started to read:

> Dear Dr. McCarthy:
> Please respond to the issues below within 60 days of receipt of this letter. Failure to receive either a full response or a request for a 30-day extension will result in a withdrawal of the submission.
> Determination: Deferred

"Look at point number three," I said. "'In the risk-level section, please change the risk level . . .'" The FDA had approved dalba in 2014 to treat skin infections, but that was before superbugs had established a real foothold in our hospital; no one knew if the drug actually worked. The IRB was concerned that we were placing patients at risk by suggesting that we could treat their infection when, perhaps, the bacteria had already become resistant. A new enzyme or efflux pump could render the drug useless, and I needed to disclose that. It was, perhaps, a fair point. "Okay," I said. I took a few deep breaths. "I'll change that."

Earlier that week, I had mentioned dalba in a faculty meeting and was met with a mix of enthusiasm and skeptical scowls. "How do we know it's any good?" a medical professor asked. "Or that it's really safe?" asked another. A seasoned physician in the back of the room noted that the FDA occasionally gets it wrong. "Does anyone remember Omniflox?" He reminded us that in 1992 the broad-spectrum antibiotic had been approved to treat skin, lung, and urinary tract infections; three months later, it was pulled after it was linked to several deaths. The drug was discovered to cause liver and kidney dysfunction, along with low blood sugar and hemolytic anemia, a condition that causes red blood cells—which deliver oxygen throughout the body—to self-destruct. "We need to look closely at this," he said, "before we expose patients." While I had addressed a number of possible limitations in my protocol, I hadn't confronted the specter of Omniflox. It was clearly an oversight.

Beyond the safety concerns, I discovered that one of the most appealing aspects of dalba—that patients could be discharged after a single infusion—was also one of the most problematic. Physicians want to keep an eye on their patients, to actually *see* them get better. The idea of giving someone an experimental drug and then ushering him or her out of the hospital seemed unwise. For my fellow doctors, decreasing length of stay was not a compelling argument. The physicians I work with want to know a new drug works before they use it, and they aren't necessarily comfortable relinquishing control. I admire them for that, and I knew that I would feel the same way.

I had seen the test tube data and was convinced that dalba would

work for our patients, but the IRB's concerns had merit. I found myself
stuck in a puzzle: I wanted to study the drug to prove it worked; others
wanted to see proof that it worked before I studied the drug. "I'm not
entirely sure how to proceed from here," I said.

"We're not the first to deal with this," Tom replied while calmly rereading the rejection letter. "Remember, the IRB isn't the enemy." He
pointed toward the door. "Disease is the enemy. Infections are our enemy." I rolled up my sleeves and let out a deep sigh. I knew the deferral
letter would set back the trial by several months. "The IRB's concerns
may be valid," he added. "Let me put on some coffee."

TOM WALSH DID NOT always want to be a doctor. When he was a
young boy in Danbury, Connecticut, he dreamed briefly of becoming a
soldier, just like his father. John Walsh had been a reconnaissance ser-
geant in the Twenty-Ninth Infantry Division and fought in World War
II alongside his childhood friends at Omaha Beach on D-Day, narrowly
escaping death. Tom grew up hearing tales of his father's maneuvers
behind German lines, gathering intelligence for the Allies in the Ardennes forest in France, and of his uncles flying B-17 bombers all over
Europe.

During his time overseas, John Walsh exchanged letters with a young
woman from Danbury. Their wartime correspondence continued until
December 24, 1944, when John's unit was attacked in a surprise air raid
in France. Shrapnel destroyed his knee, but it earned Walsh a Purple
Heart and, like Gerhard Domagk, a reprieve from further active duty.
The young soldier returned home and married the woman; not long after that, Tom came along.

A sense of patriotism and service had been instilled in Tom Walsh
from a young age—he often spoke of our antibiotic work as "the mission"—
but life took an unexpected turn when he was seven years old. One day
after school, Tom learned that his mother wasn't feeling well. The next
morning, the aches were more intense, and she was sent to the hospital.
His mother was diagnosed with an aggressive form of cancer, gastric

carcinoma, and one year later, at the age of thirty-three, she was dead. "One day my mother went to the hospital," he told me, "and never came home." That's when Tom decided to become a doctor.

When he was a boy, Tom saw that physicians and other health care workers possessed a remarkable quality, one that he was not sure he could find in the military: compassion. After his mother died, his father was hit with a devastating hospital bill. "Something around a hundred thousand dollars in today's money," he recalls, "and there's no way we could've paid it." The small-town doctors who treated his mother tore up the bill. "They just forgave it." The physicians couldn't save Tom's mother, but they saved his father.

The generosity inspired Tom to pursue medicine, but the familial pull of the military was difficult to shake. He was a scholar-athlete in his blue-collar town, excelling in the sciences as well as track and field, and his family doctor nominated him for the US Naval Academy. Walsh had trouble saying no (he still does) and would have become a midshipman if it weren't for an errant comment during his academic interview: while sketching out the details of his young life, Tom confessed that his true passion was medicine.

The statement derailed his appointment—the interviewer told him, "We train sailors and marines, not doctors"—so Walsh opted for Assumption College, a small Catholic school in Worcester, Massachusetts, where he was given a full scholarship to study biology and chemistry. He breezed through college and then enrolled in medical school at Johns Hopkins University before entering a sequence of increasingly specialized training programs—internal medicine, microbiology, pediatrics, infectious diseases, hematology/oncology—that led to his current position at NewYork-Presbyterian Hospital.

The trajectory of Tom Walsh's groundbreaking research career can be traced back to a doctor named Bernadine Healy. They met when he was a medical student intent on curing cancer, and she was a rising star in the field of cardiology, with lavish research support and a gift for writing. Together they cared for a patient with a rare condition: a yeast infection of the heart. The formal diagnosis was *Candida* endocarditis, and it perplexed them both. The cause of the fungal infection was difficult to pinpoint and even more challenging to treat. One evening, after rounds had concluded for the day, Healy invited Tom to work in her laboratory. There he realized that research was his calling. She would go on to become the first female director of the National Institutes of Health, and he would eventually change the way doctors treat infections.

Along the way, Tom had a couple of kids—their drawings are strewn across his office—and he still enjoys offering parenting advice. (When my daughter was born, he bought her a copy of his favorite children's book, *The Little Engine That Could*.) These days, he spends many late nights alone in his office, away from his family, toiling away on a manuscript or grant proposal. On a typical morning, as he goes for his predawn jog along the East River, I begin reviewing the overnight text messages I've received from him—"Check out the new trial with eravacycline!"—along with his forwarded email correspondences. Although I have never told him this, it's one of the highlights of my day.

Like Fleming and Domagk before him, Dr. Walsh uses military parlance. He routinely speaks of tactical maneuvers, corps values, and "the enemy," and sets appointments according to military time. A conversation about antibiotics can easily veer into a dissection of World War II general George Patton's tactical genius, replete with diagrams and references. In the early years of our working relationship, I was the aide-de-camp to his commanding officer. He dispensed orders, and I followed them. Now he likes to say that we are a team, but I know who the captain is.

He does not live in an extravagant home, nor does he drive an expensive car. His is a Spartan existence, constantly on the move. (Keeping up with Tom on foot can be just as challenging as following his boundless thoughts.) He finds value in human connection, and his approach is one that has inspired a generation of young doctors.

Tom gives out his cell phone number to anyone who asks for it (and those who don't), and he has an open-door policy, which means that our meetings are constantly interrupted by physicians seeking his wisdom. To be around him is maddening—it's difficult for him to complete a thought without being interrupted by the phone, email, or a knock at the door—but it's also enchanting. On his desk, next to piles of reminders and receipts and flight itineraries, sits a small note with a phrase that marries his medical and military mind-sets: "We defend the defenseless." It's his guiding principle—a through line that unifies all of his disparate projects and responsibilities—and it reminds me of the motto Gerhard Domagk created after the Great War: "Whatever contributes to the preservation of life is good; all that destroys life is evil." In the first few years of working with Tom, I searched for my own mantra or saying, but nothing quite fit. A more senior doctor once gave me a pin that said, "Walk the Dog," to serve as a reminder that it was okay to spend a few minutes of the busy workday doing something wholly unrelated to patient care, but it wasn't the thing I was looking for. I don't even like dogs.

Tom is calm under pressure, and, unlike me, he is often able to take the long view, putting negative feedback and frank criticism in perspective. He can shake off things in a way I simply can't. I'm a ballplayer at heart, drawn to the thrill and immediacy of competition, and I have trouble with unknowns. I'm not suited for deferral letters. I see things in absolutes—accept or reject, win or lose—where Tom sees shades of grey. He sets aside his own interests for the sake of *the mission* and thus was the ideal person to help me absorb the deferral letter from the IRB.

Tom returned with a pot of coffee and put an arm on my shoulder. "Some of their concerns are valid," he said again. "Many of them."

"Like what?" I began to read from the letter.

CHAPTER 8

Oversight

A MAXIM IN MEDICINE is that antibiotic resistance comes with a fitness cost, meaning that when bacteria become impervious to antibiotics—when they mutate into superbugs—they sacrifice something vital in return. Devoting resources to evasion leaves superbugs exhausted and unable to spread. It's a phenomenon that infectious disease specialists count on, but it turns out this paradigm is changing: superbugs have recently become more fit *and* more virulent. In other words, they're getting smarter and stronger.

This had profound implications for my dalba trial and the risk associated with participation. It was clear from the IRB's terse wording that I had underestimated the possible dangers dalba posed to patients. I was offering a false sense of security by telling them that I could potentially cure their infection and shorten their hospital stay. But it was far from certain that this would, in fact, be the case. I hadn't mentioned efflux pumps—the microscopic vacuum cleaners that bacteria use to suck up and expel antibiotics—or any of the other chemical modifications that they might use to neutralize dalba. I hadn't mentioned that bacteria were becoming more aggressive and that my drug might not work. The protocol was in need of a drastic rewrite.

To gain a bit of perspective, I reached out to several experts to understand how they approach clinical trials and antibiotic research. I started with Brad Spellberg, chief medical officer of LAC+USC Medical Center, a top-flight, oddly punctuated health care and research center. Spellberg is a thoughtful and devoted physician-scientist; he's also a provocateur. At a major conference in San Diego, I listened with delight as he stood at a podium, calling out pharmaceutical companies by name for the trials they should have done but were scared to attempt. (He told the standing-room-only audience that Allergan lacked the "cojones" to conduct a bloodstream-infection study with one of its drugs.)

Spellberg and his colleagues believe that resistance already exists to all antibiotics, including those *we have not yet discovered*. To understand how this is possible, we might invoke the infinite monkey theorem, which argues that a monkey hitting keys at random on a computer keyboard for an infinite amount of time will eventually produce coherent text, including the complete works of William Shakespeare. By way of comparison, microbes are constantly mutating, hitting the proverbial keys in novel combinations, and those sequences produce enzymes and pumps that can deflect or destroy any antibiotic. Spellberg and his team have noted that antibiotic resistance has even been discovered "among bacteria found in underground caves that had been geologically isolated from the surface of the planet for 4 million years." It's a terrifying thought that called into question the very essence of my trial. I reached out to Spellberg because I valued his skepticism, and I thought he might give me the most critical eye.

"There are already widespread resistance mechanisms in nature to drugs we haven't invented yet," he told me one morning before rounds. "When we come out with a new antibiotic, people think new mutations occur after we start using the drug, but that is false. The much bigger problem is that there are low levels of preexisting resistance mechanisms that we can't yet detect. When we dump a new antibiotic into the environment, we apply selective pressure and resistance grows." Eventually we will run out of new drug targets. "We need to be smart about this," he added. "Bacteria use antibiotics judiciously. Humans do not."

Spellberg told me that the solution is to take the long view. "We don't want a flood of new antibiotics," he said. "We need a slow and steady drip." Bringing a number of antibiotics to market simultaneously would be problematic, he explained, because resistance would occur in tandem. We desperately need more antibiotics, but it would be a mistake to test all of the best candidates simultaneously.

After surveying a handful of experts, including some who requested anonymity because of their relationships with Big Pharma, I revised the dalba protocol, conceding that the risk had been understated, and resubmitted it. "Fingers crossed," I said to Tom. The leitmotif of his expansive career had been to solve the unsolvable; I had faith that together we could steer our study through the latticework of approvals and regulations. "I feel pretty good about this."

"Now we wait," he replied.

I went back to seeing patients, and Tom returned to writing grant proposals. What struck me in the weeks that followed, as we waited for a response from the IRB, was the rising number of patients who were admitted to my hospital because oral antibiotics were no longer working. They had routine infections—pneumonia or urinary tract infections—that in prior years could have been treated at home with a week's worth of pills. But the treatments simply weren't strong enough. Bacteria really were getting smarter and stronger. In the week after I revised the protocol, Jackson passed in and out of my emergency room twice. He told me the infection prevented him from seeing his daughter's dance recital and his son's first basketball game. "Nothing seems to work," he said. And he was right. He was coping with a chronic infection and hoped that he wasn't spreading it to others.

This shift in the way we treat infections—from oral to intravenous antibiotics—was contributing to a burgeoning crisis at the hospital. Due to overcrowding, patients were waiting up to thirty hours in the ER just for a bed to open up. On some days, we had to turn ambulances away. There simply wasn't the space for the additional bodies, and patients were instructed to look elsewhere. Jackson was just one of hundreds of patients I've cared for with a superbug infection. Many of these people died, and even more were left profoundly debilitated. One woman, a fifty-nine-year-old receptionist from Staten Island, told me that she knew a recurring spinal infection wouldn't kill her, but she wished that it would. "I'm tired of playing this game," she said, and noted that she now spent far more time in the emergency room than in her own apartment. "Enough is enough."

There was no good way to predict who would contract an infection or who would succumb to the illness. We were all at risk because bacteria don't discriminate—they attack all comers: the young, the old, and everyone in between. They were outfoxing us, and in some ways it felt like we were returning to a pre-antibiotic era, one in which a century of scientific progress had simply been erased. While waiting for a response from the IRB, I kept asking myself: *Why is it so hard to make a new antibiotic?*

To APPRECIATE THE CHALLENGE that antibiotic developers face, it's useful to know a bit about the origin of medical oversight in the United States. This backdrop provides the basis for FDA's substantial regulatory powers, and it helps explain why antibiotic approval is so frustratingly slow. More importantly, it reveals what might be done to fix the problem.

What we know as the FDA—a sprawling bureaucracy with a $5 billion annual budget—once looked radically different. The agency began in the nineteenth century with just a few scientists at the Bureau of Chemistry in the Department of Agriculture performing lab tests and issuing reports on the quality of foods and drugs. With the extraordinary rise of free trade in the early twentieth century, however, its mission changed.

As the link between unrestricted commerce, unsanitary working conditions, and public health was established, an outraged citizenry compelled Congress to protect the food supply. Just a few months after Upton Sinclair published his grim novel *The Jungle* in 1906, President Theodore Roosevelt signed into law the Pure Food and Drug Act, which prohibited the interstate transport of food or drugs that had been "adulterated" or mislabeled. With the stroke of a pen, that small group of government scientists was given remarkable power. The chemists were now charged with regulating a landscape where companies routinely misrepresented their products to maximize profits. It was common for wholesalers to alter spoiled produce (copper sulfate can make rotten vegetables appear ripe) and to enhance the color and smell of medications. Embalming fluid was added to milk to keep it from going bad. Before 1906, none of these practices were regulated, and it led to thousands of deaths for which no one was held accountable.

In the years following the passage of the law, as pharmaceutical companies refined the means of production and mastered the art of marketing and mass distribution, a stream of new products came to market, including stimulants, powerful painkillers, and drugs to treat cancer. Government oversight evolved in concert with these developments—the agency was formally named the US Food and Drug Administration in 1927—but there was a problem: it did not have the power to ensure that these drugs were safe *before* they were given to patients.

In the fall of 1937, Gerhard Domagk's prized antibiotic, sulfanilamide, entered the US market shortly after it was used to cure Franklin D. Roosevelt Jr., son of the thirty-second president, of a sinus infection. A salesman for the S. E. Massengill Co. in Bristol, Tennessee, reported a demand in southern states for the drug in liquid form to treat strep throat, and the company's lead chemist found that sulfanilamide would dissolve easily in a sweet liquid called diethylene glycol. Massengill performed some preliminary tests for flavor and fragrance and shipped the elixir to Tennessee. If it worked for a Roosevelt, it would presumably work for others, too.

The drug had not been tested for toxicity, however, because there was no requirement to do so. Unsuspecting doctors and patients had no idea that this remarkable antibiotic had been mixed with antifreeze. By the time the government got wind of what was happening, its intervention was sluggish and ineffective. Many of the 107 Americans who died were children. Massengill was fined $16,800, the largest sum levied by the FDA up to that point.

The sulfanilamide tragedy underscored how little the FDA could actually do. The following year, President Franklin Delano Roosevelt signed the Food, Drug, and Cosmetic Act of 1938, increasing federal authority over drugs by mandating a *pre*market safety review. It gave the FDA the power to ban phony labeling, recall ineffective products, and designate certain drugs as safe for use only under the supervision of a doctor. It allowed the FDA to define what is safe and what isn't, a power the agency still holds today. All of this occurred because outraged citizens demanded change.

The FDA began removing thousands of dubious products from the market while overseeing the approval of a staggering portfolio of safe and effective new ones. By the early 1950s, 90 percent of the prescriptions filled by patients were for drugs that had not even existed in 1938. More effective therapies were developed in that short window than in all of previous human history, and regulators scrambled to keep up. The FDA's offices rapidly expanded and its budget swelled as top scientists were recruited from the public and private sectors to support the golden era of antibiotic development.

One of those new hires was Frances Oldham Kelsey, a physician with a doctoral degree in pharmacology. In 1960 she was asked to review an FDA application for a tranquilizer that had become popular in Europe for its effectiveness in treating morning sickness. A pharmaceutical company in Cincinnati wanted to sell this new drug in the United States, and Dr. Kelsey was asked to review the regulatory documents. Thalidomide really did improve some symptoms in pregnant women, but unbeknownst to doctors and patients, it had a chemical property that allowed it to cross the placenta. That feature would eventually lead to thousands of birth defects around the world, including phocomelia, a malformation of limbs.

For nearly two years, Dr. Kelsey refused to approve thalidomide in the United States while the drug's manufacturer publicly attacked her. (She was no stranger to slights; Dr. Kelsey had been accepted into a doctoral program at the University of Chicago because, with her gender-ambiguous name and short hair, brushed back severely, researchers assumed initially that she was a man.) Despite its widespread use in

*Dr. Frances Oldham Kelsey
speaking to a Senate government
operations subcommittee.*

other countries, Kelsey sensed that something was wrong with the drug, and she insisted that more testing was necessary before it could be given to patients. The thalidomide tragedy was largely averted in the United States because of her careful work and her refusal to bend to public pressure. John F. Kennedy feted her with the President's Award for Distinguished Federal Civilian Service, and she spent the remainder of her life protecting consumers at the FDA. She worked there until she was ninety.

Whenever I'm frustrated by the slow pace of drug approval, I remind myself of Frances Oldham Kelsey.

TODAY GOVERNMENT watchdogs oversee products made by over one hundred thousand businesses, and every year several hundred toxic drugs and faulty medical devices are taken off the market. The agency also weighs in on medical controversies: the FDA recently announced that there was insufficient evidence to recommend over-the-counter antibacterial soaps over washing with plain soap and water. The finding was a surprise, but the dictum worked its way into my hospital as doctors and nurses altered their hand hygiene to comply with the recommendation. When the FDA speaks, we listen.

The scope of this work is incredible—the number of lives saved by the FDA likely rivals the number saved by penicillin—and in a given year it regulates more than $1 trillion in commercial goods, including $275 billion in pharmaceuticals. A deferral or rejection letter from the FDA can send a stock into free fall.

There have been some steps backward, however. In 1994, Congress limited the FDA's ability to regulate herbal remedies and supplements, which has led to countless overdoses and deaths. In my own practice, I have seen many patients who shunned Western medicine in favor of herbal options, only to return to the hospital once those remedies failed or proved too toxic to tolerate. One woman told me she had refused treatment for her early-stage breast cancer because everyone she knew who received chemotherapy had died. She opted for herbal potions and came back to my hospital after the cancer had spread. Her husband and I watched her die from a treatable form of cancer, an empty container of supplements on a nightstand next to her bed. That same week, I cared for a man who traded his insulin for a dietary supplement he'd purchased online; two weeks later, he returned to my emergency room unconscious, in the throes of a diabetic coma, with blood sugars so high they didn't register. I fumbled for words when his wife asked me how something like this could happen.

The FDA can do many things, but it cannot regulate drug pricing. If a company wants to price gouge, as Turing Pharmaceuticals did in 2015 when it acquired the rights to the antiparasitic drug pyrimethamine and raised the fee 5,000 percent, the agency is powerless to stop it. Other companies have adopted this strategy, as well as some unexpected talking points. In 2018, the president of Nostrum Pharmaceuticals, Nirmal Mulye, told the *Financial Times* that he had an ethical obligation to raise the price of the antibiotic nitrofurantoin by 400 percent. The drug is one of the most commonly prescribed treatments for lower urinary tract infections—it's on the World Health Organization's list of essential medicines—and its price jumped from $474.75 to more than $2,300 per bottle overnight. "I think it is a moral requirement," Mulye said, "to make money when you can."

The agency that was once commended for its ability to evolve and protect is now having trouble keeping up. The FDA has been accused of dragging its feet, of being unresponsive to the needs of dying patients, and physicians often feel the approval process is disconnected from their everyday experiences. It's telling that my colleagues were averse to using dalba despite the fact that it had received FDA approval. The agency's studies might say it's safe, but that meant relatively little in the eyes of physicians who were already anxious about treating superbugs.

To accelerate and improve the process, the FDA created a "breakthrough therapy" designation for the best investigational drugs. Once a treatment receives this label, the FDA commits to working closely with the drugmaker to devise the most efficient way to generate the evidence needed for approval. A breakthrough drug must treat a serious or life-threatening condition and preliminary clinical evidence must demonstrate "substantial improvement" over existing therapies. Antibiotics rarely qualify.

One of the few that did was dalba. But it's been a struggle to find more antibiotics that make the cut. It's actually quite hard to show that a new antibiotic is better than the dwindling stockpile that we've already got, and bacteria keep figuring out ways to inactivate what we throw at them.

Once an antibiotic is cleared for use by the FDA, there is no guarantee that patients will have access to it. My hospital occasionally faces shortages of crucial treatments; there are times when I'm unable to give AIDS patients the proper medications for a certain form of fungal pneumonia. "We need a different drug," I find myself saying to anxious patients and the surrounding medical students. "Something else—anything else. What are our options?" It was a shock, at first, to offer an inferior treatment at my world-class hospital, but the shock has since worn off. I know that on some days lifesaving drugs are simply unavailable.

On January 18, 2013, the FDA reported a shortage of doxycycline, the drug used to treat Lyme disease and a number of other tick-borne illnesses, as well as cellulitis and the bacterial infection MRSA (methicillin-resistant *Staphylococcus aureus*. The problem was attributed to increased demand and manufacturing issues, and it put countless lives in danger. Someone was letting them down, but I couldn't figure out who was responsible. The price of doxycycline soon skyrocketed from 6 cents per pill to $3.36, an increase of more than 5,000 percent. It reminded me of the financial crisis of 2008: there was plenty of finger-pointing, but no one personally took the blame.

Two years after the doxycycline shortage, I was working at my hospital's satellite location down in Chinatown when I received a notice about a shortage of piperacillin-tazobactam, a combination antibiotic that I have used to treat hundreds if not thousands of patients with pneumonia, urinary tract infections, and colitis. It's one of the most reliable drugs I've ever used, and suddenly it was about to disappear. I was never told why this happened; I just accepted it and moved on. But ignoring the problem is becoming more difficult. Between 2001 and 2013, there were 148 shortages of antibiotics, and doctors across the country resorted to second-class treatment options. Most patients didn't even know it was happening. That volatility also destabilized investment in new drugs, providing cover for CEOs who want to maximize profits through morally dubious mandates.

It's hard to believe that two generations after penicillin first hit the market, and after two hundred million lives saved, there is now a worldwide shortfall. But that's the reality we now inhabit. Pfizer, the sole manufacturer of penicillin G benzathine in the United States, has attributed the problem to manufacturing delays, but the real answer is more nuanced. Only four companies produce the active ingredient in penicillin, and those manufacturers, based in China and Austria, keep production levels low because the drug offers so little profit.

Companies don't bother to keep up because penicillin is mostly used to treat diseases that affect poorer countries—US doctors don't see much rheumatic heart disease anymore, but Indian doctors do—and demand simply outstrips the supply. "There is a market failure in the penicillin sector," Dr. Ganesan Karthikeyan, a cardiologist in New Delhi, told Al Jazeera. "There is a demand, but it comes from the poor." One pharmaceutical executive told me it costs around $20 million per year to maintain a factory to produce penicillin, and suggested that instead of investing in new drugs, we should ensure proper access to the ones we already have.

Countries with limited purchasing power are the ones who need Fleming's discovery the most, yet they are having the most trouble acquiring it. We continue to focus on proposals to spur the development of antibiotics through tax incentives and patent extensions and even an options market—but we're ignoring the challenge of maintaining adequate production of cheap drugs that are proven to work. The FDA's process may seem slow, but once an antibiotic is approved, we must ensure that it remains available to the people who need it most. Heart disease from untreated strep throat is entirely preventable—the infection can be cured with a short course of penicillin or other inexpensive antibiotics—yet it kills more than three hundred thousand people worldwide every single year.

CHAPTER 9

Backwater

"STILL NOT SATISFIED," I said to Tom several weeks later. "Modifications required." After reviewing my revised protocol, the board had a different set of questions for us: How would dalba be paid for? Was Allergan supplying the drug? How many patients would take part in the study?

"Who are these people?" I asked. "Have they ever taken care of a patient? Have they ever seen someone die because—"

"Deep breath," Tom said. "Just answer the questions and resubmit." I shook my head in disbelief.

"Adapt," he added. "Improvise, adapt, and overcome." It was the motto of the United States Marine Corps, and one of his mantras—I'd heard him say it to other doctors and to his children—but it hadn't quite sunk in yet. I've always wanted quick results. In many ways, I was poorly equipped to handle the glacial pace of clinical research. I followed Tom's advice and responded to the questions—yes, Allergan was supplying the drug, and we expected to recruit around one hundred volunteers—and then I went back to seeing patients.

Writing *one hundred* was a reminder of the scope of my trial. The biotech company Achaogen had to interview and evaluate 659 patients to enroll just 14 in its study of a superbug. I might have to vet thousands of patients to find enough volunteers for mine. I practiced my pitch in the mirror, straightening my tie as I spoke about the potential benefits of an untested antibiotic—one that might be as dangerous as Omniflox. "We have every reason to believe it will work," I said to my reflection. "We believe it *will* work," I added. "It will work, I believe."

The next few weeks were difficult. I wanted to take on the responsibility of a trial, and I wanted to know if dalba actually helped people, but maybe it wouldn't. Even if a study is approved, it doesn't always run until completion. Sometimes trials are shut down because a drug is deemed to be ineffective or dangerous halfway through the study—or even earlier. The FDA keeps tabs on drugs even after they're approved because the agency knows that danger can strike at any moment. An independent group monitors data as it is collected, and at any point, a principal investigator can receive a call explaining that a trial must be halted. It's an uncomfortable thought—it means that someone else knows about the futility of my trial before I do—and I often imagine how I would handle that call. What would I say to the colleagues and patients who trusted me?

Around the time that I received the deferral letter, the United Nations held its first General Assembly meeting on drug-resistant bacteria. It was only the fourth time the General Assembly had taken up a medical issue (the others were in response to HIV, Ebola, and noncommunicable disease) and I paid close attention to how people from other countries spoke about the issue of superbugs. I wanted to know if people in power truly understood what was happening.

It was clear from the outset that the aim was to alarm. "If we fail to address this problem quickly and comprehensively," UN secretary-general Ban Ki-moon said at the meeting, "antimicrobial resistance will make providing high-quality universal health care coverage more difficult if not impossible." All 193 UN member states agreed to channel the global response to superbugs along a similar path to the one used to address climate change. The nonbinding agreement would strengthen regulation, encourage innovation in drug development, and improve reporting systems that monitor how antibiotics are used. It felt like a step in the right direction, but it was so far removed from my day-to-day experience in the hospital that I didn't know what to make of it. Whether a nonbinding agreement would translate into meaningful change was anyone's guess.

After the UN General Assembly, the World Health Organization (WHO) released a list of the most dangerous superbugs, dividing the bacteria into groups—medium, high, and critical—based on prevalence, level of resistance to treatment, and mortality rates. The list included familiar pathogens, including *Staphylococcus aureus*, *Streptococcus pneumoniae*, and *Enterococcus faecium*, which had become resistant to standard therapy, as well as more obscure bacteria, such as carbapenem-resistant *Acinetobacter baumannii*. (My patient Jackson had acquired half of the bugs on the WHO's list.) The accompanying press clippings used an alphabet soup of acronyms to refer to these organisms—MRSA, VRE (vancomycin-resistant enterococci), VRSA (vancomycin-resistant *Staphylococcus aureus*), and CRE (carbapenem-resistant *Enterobacteriaceae*)—which I hoped would make the complicated organisms seem more accessible.

The one I'm most interested in is methicillin-resistant *Staphylococcus aureus*, perhaps the most famous superbug. MRSA used to reside in very specific spots—gymnasiums, health care facilities, and locker rooms—which meant that only certain people would come into contact with it. For many years, most cases of MRSA affected athletes and the elderly. But in the late 1990s, MRSA started seeping into the community, putting just about everyone at risk.

It took doctors a few years to appreciate the magnitude of this shift. In the early 2000s, MRSA was often missed by physicians who simply didn't think to look for it. This led to improperly treated infections, which prompted even greater antibiotic resistance. We now know that MRSA likes to live in dairy cows, which serve as a reservoir for the bacterium, allowing it to pass back and forth between species. Dalba is effective against MRSA and could curb its spread both in the hospital and in the community, but without IRB approval, I couldn't use it to treat MRSA or any other infection. It just sat on a shelf somewhere, far removed from the patients who might need it.

OVER THE NEXT FIVE MONTHS, I received three more deferral letters from the IRB. My rewrites simply weren't helping. If anything, I was making the protocol worse with every revision. After the fifth deferral arrived, I received a call from a researcher at Allergan. The company was thinking of canceling the study. "If you can't get it approved," she said, "what's the point?" My year of personal and professional discovery had turned into annus horribilis.

As the days and weeks passed, my frustration built. I knew some patients weren't receiving optimal medical care; bacteria were evolving, infections were becoming more aggressive, and I was doing nothing to move the ball forward. What started as a small rash on the leg might spiral to the heart, bone, and brain, leading to a prolonged hospitalization, months of rehabilitation, and occasionally amputation. I was prescribing the best drugs I had, but that was no longer good enough.

My patients were increasingly whisked away to the operating room for surgical removal of gangrenous limbs because antibiotics were no longer effective. I looked on helplessly as a skin infection ravaged a young man's body, shutting down his organs one by one. In the bed next to him, a marine officer was battling necrotizing fasciitis, a flesh-eating disease that developed after he popped a pimple near his groin. As I broke the news to his daughter that he might never walk again, I fought off the urge to scream. Something was very wrong.

I had been drawn to medicine because, in college, I heard that working in a hospital was like being on a team. Moving from the pitcher's mound to morning rounds felt like a natural transition, one that would provide me with a new roster of teammates. At first, this was true. Medical school was actually less stressful and more enjoyable than my premed experience at Yale. But as I progressed through the ranks of residency and subspecialty training, the team dwindled. Fewer people possessed my clinical interests and expertise, and I came to feel isolated. It was often just a patient and me, alone in a room, discussing a series of terrible options. It felt a bit like pitching with a weighted baseball, one that I knew I couldn't throw for a strike.

In meetings, I was distracted. My mind often wandered to these men and women, the ones who were quietly deteriorating. The IRB had promptly approved many of my other studies. What was the holdup now? What was it about dalba? Anger sluiced over me every time I thought about the delays. Maybe it was the affiliation with Allergan. Yes, the company had a good reputation for aggressively investing in antibiotics, but it was also known for curious moves that bordered on corporate trickery. It had transferred the patents for its blockbuster eye drug, Restasis, to the St. Regis Mohawk tribe in upstate New York, allowing the company to invoke tribal sovereign immunity to fend off patent challenges from generic drugmakers. I was torn by the tactic—it seemed to exploit a group that had already been so exhaustively exploited and would limit public access to cheaper generic drugs—but it was going to create a revenue stream for the tribe while freeing up more resources for Allergan to invest in antibiotic development. (The transfer was later thrown out in court.)

"Hang in there," Tom said day after day. "Just stay focused and hang in there." I knew he was holding back, trying not to step in, so that I would learn how to do this on my own. At times, our exchanges could feel like outtakes from *The Karate Kid*. He was my Mr. Miyagi, an enigmatic guru helping me to tap into something from within, making a point with a gentle nod or by asking me a series of questions. It was a difficult way to learn.

I thought of his advice day after day as I edited the protocol, and the children's book he'd given me, *The Little Engine That Could*, to keep me going. After six months of revisions, my study had morphed into something different from the one I had initially submitted. My colleagues offered their sympathy, telling me of regulatory delays they had experienced, but this was something altogether different. It usually took around 130 days to get a trial off the ground. I was on pace to double that.

I reviewed the deferral letters over and over again, looking for any clue that the IRB might simply shut down the study before it ever got started. I revised the wording on the trial's consent form and walked over to Tom's office to resubmit the protocol. "I'm losing my mind," I said. Sitting there, surrounded by framed manuscripts and textbooks, I felt a mix of hope and despair. He was at the pinnacle of our field, a career marked by innovation and, I gathered, occasional setbacks. But he retained an eternal sense of optimism.

Shortly after I hit Send, Tom received a phone call. I moved to step out of the room, but he waved me toward him, indicating that I should stay. As the person on the other end spoke, he closed his eyes and wrinkled his brow. A moment later, he turned on the speakerphone. The caller was a pediatric infectious diseases specialist from Denver. "I have a fifteen-year-old girl," the man said, "and she's dying."

I recognized the hint of terror in his voice. Tom clasped his hands on the conference table. "How can we help?" he asked.

"She's got a mold infection, something called *Scopulariopsis*."

Tom looked at me and shook his head. "Tell me about her." He ran his thumb and forefinger along the length of his chartreuse necktie as the man spoke.

"I've tried just about everything," the specialist said, "except calling you."

Tom was one of a select few people in the world who knew how to treat *Scopulariopsis*, and perhaps the only one in North America. He pulled a pen from his coat pocket and began to take notes. For twenty minutes, we listened as the doctor from Denver relayed the details of the case. Every few minutes, he would interrupt his narrative to remind us that he was searching for help. Much of the attention surrounding superbugs has focused on bacteria, but drug-resistant fungi can be just as deadly. Most companies had shied away from investing in development because fungal infections are relatively rare and profit margins are slim. That fact was of little use to the team in Denver.

"Here's what you need to do," Tom said. "First, double the dose of posaconazole [an antifungal]. She can tolerate much higher levels." He proceeded to rattle off a series of diagnostic tests that would allow the doctors to monitor her progress, closing his eyes for long stretches as he spoke.

"Got it," the physician said. "Thank you so much. What about terbinafine?"

Tom shook his head. "Absolutely not. She may also need a granulocyte transfusion. Have you given one before?" Tom was suggesting a white blood cell transfusion in a last-ditch attempt to bolster the girl's immune system. He waited for a response, but there was none. Tom looked at his wristwatch and then at his calendar, which was plastered across two computer monitors on his desk, and then at me. I knew what he was thinking, and began to search on my iPhone for direct flights to Denver.

The once-porous boundary between Tom's personal and professional lives had dissolved years ago. Now, it seemed, every waking moment was devoted to helping vulnerable patients, developing drugs, and pushing the margins of what was possible to accomplish in a single day. The professional was deeply personal, and in his presence, none of it felt like work. It was, simply put, his calling.

"This is very important," Tom went on. "She's pancytopenic, and her body may not respond to any of this. Everything I've said may be for naught." The child's bone marrow had malfunctioned; it was no longer producing red blood cells, platelets, or white blood cells that typically fend off infection. This condition, pancytopenia, left her exposed to unimaginable infectious agents, including the mold in her basement.

A deep sigh came through the speaker. "Then what?"

"Emergency stem cell transplant," Tom said. This would replenish the patient's bone marrow with fresh cells—stem cells—that could transform into white blood cells and restore her immune system.

The caller paused. "Really?"

Tom closed his notebook and nodded at me. Twenty minutes later, he was in a taxi bound for John F. Kennedy Airport.

FUNGAL WORK sometimes feels like an intellectual backwater. Bacteria get all of the attention, both from academics and from industry, and the world's small cohort of fungal specialists, known as mycologists, shrinks a bit more every year. There's nothing sexy about being an expert in yeast infections, but when disaster strikes, we need people like Tom Walsh to pick up the phone.

Fleming's chance discovery of the first antibiotic is enough to spark the imagination of any child with a budding interest in science, but the discovery of the first antifungal drug is equally compelling. It's a story about two brilliant women that has been omitted from most science books, and it's unknown to most of today's young physicians.

Elizabeth Hazen was orphaned when she was just three. She spent the majority of her childhood bouncing around rural Mississippi at the turn of the twentieth century, living first with her grandmother and then with her uncle. After high school, she attended what is now known as the Mississippi University for Women and then moved to New York to study bacteriology at Columbia University. Her studies were interrupted by World War I, during which she served in the army, but she eventually obtained a doctorate before moving to New York City's Division of Laboratories and Research in 1931.

A dozen years later, during World War II, physicians noticed that penicillin was protecting soldiers from bacterial infections, but many were contracting fungal diseases. There was no cure, and some suspected that Fleming's discovery was actually predisposing the men to the fatal infections. Elizabeth Hazen was tasked with finding a way to treat them. In her laboratory, she went about the painstaking project of isolating organisms found in soil samples and testing them against two fungi that were known to infect humans, *Candida albicans* and *Cryptococcus neoformans*. When Hazen had a potential hit, she mailed her samples in a mason jar to a chemist in Albany, New York, named Rachel Brown.

Upstate, Dr. Brown would purify the samples and send the drugs back to Hazen for testing in animals. Their work progressed at a blistering pace, and was made possible by the remarkable efficiency of the US Postal Service. In just a few years, they reviewed thousands of molecules, but almost all the drugs that killed fungi in test tubes turned out to be highly toxic in animals. Finally, after years of searching, one worked. The compound destroyed fungi without harming animals or humans, and, of all places, it had been found in the garden of Hazen's friend, Jessie Nourse. A bacterium in the soil was producing the antifungal drug; Hazen named it after her friend, *Streptomyces noursei*.

The two researchers announced their results at the New York meeting of the National Academy of Sciences in 1950 and immediately attracted interest from Big Pharma, which was entering its golden age. Hazen and Brown were suddenly rich and, briefly, famous. They invested their millions in a nonprofit that funded more research. They named their fungal drug nystatin," after the NY State Department of Health, and continued to collaborate throughout their lives, discovering two more antibiotics together.

Nystatin has saved countless lives—I prescribe it all the time—and it's even used occasionally to restore damaged artwork. (After a flood in Florence, Italy, curators at Boboli Gardens sprayed nystatin on more than two hundred paintings to protect them from mold.) Its effect has been staggering: it is on the World Health Organization's list of essential medicines and is one of the cheapest and most effective products on the market today. But the nystatin story isn't taught to fledgling scientists, medical students, or residents. Forgetting this piece of history speaks to our failure as educators. Everyone knows about Alexander Fleming, but no one knows about Elizabeth Hazen and Rachel Brown.

AS THE DAYS TICKED BY, and I continued to wrestle with the IRB, economists in the leading medical journals argued for changes to sweeten the deal for pharmaceutical companies in the form of more drastic push and pull incentives. (The former lower the cost to develop a drug; the latter reward success.) Antibiotics are priced relatively lower than drugs in other fields, such as oncology and rheumatology, and those costs typically don't reflect the curative benefit: for example, penicillin has saved two hundred million lives but costs only a few dollars. By contrast, some chemotherapy drugs are priced in the tens of thousands of dollars and extend life by only a few weeks. A number of new proposals are on the table to address this odd discrepancy.

One option is a two-tier pricing system based on diagnosis, under which antibiotics are priced lower for an empiric treatment (say, if a doctor suspects pneumonia, based on his expertise and the patient's symptoms, but hasn't confirmed it with a chest X-ray), and at a premium once the diagnosis is confirmed. This, of course, creates a disincentive to obtain a definitive diagnosis; doctors might make a guess because it's cheaper. There's also the market entry reward model, which pays hundreds of millions of dollars to any company that creates a drug that satisfies predefined public health priorities. A drug capable of killing superbugs would be worth billions, regardless of how often it was prescribed.

The most provocative idea is an options market, in which investors would have the right to buy a specified amount of an unapproved antibiotic for a fixed price. The cost would be low if the options were purchased early in development to reflect the risk to investors, since most new drugs fail, and higher later in the approval process. The owner could then sell the options to governments, hospitals, or patients for a profit. Free market proponents are big fans of this approach, but it would require something that isn't easy to come by: an open exchange of scientific information between drug developers and options purchasers. It also would adversely affect patients in low-income countries. If options are priced too high, option holders could price gouge. The FDA has also revealed that it's considering a subscription-based plan that would require hospitals to pay a flat rate for access to a specific number of new antibiotics. Medical centers would get drugs the way you and I access music and movies through online services such as Netflix.

None of these ideas seems destined to succeed on its own; perhaps a combination would work. But as I heard time and again, antibiotics simply weren't profitable—one economist told me that a great way to squander $30 million is to invest in an antibiotic—and that wasn't going to change, no matter the incentives. Beyond that, the regulatory framework and approval pathway had become so onerous that it could be difficult to get any antibiotic study off the ground. I tried to stay positive, but the daily grind at the hospital was making that difficult.

I missed Tom. He was a font of wisdom and unbridled optimism in a field studded with roadblocks. My optimism was fading, and I needed a jolt of his ever-present enthusiasm. Not long after he returned to Manhattan, I strolled over to his office to discuss a new project, one involving a new treatment for urinary tract infections. I learned that the girl in Denver was going to survive and would not need a stem cell transplant. While we were chatting, I received a short email from the IRB:

The protocol and its relevant documents stand approved.

After years of planning and months of waiting, I was finally ready to begin the dalba study—and it couldn't have happened at a better time. It was the middle of July and there is a seasonality to skin infections: hospitals see a sharp spike in cases over the summer. My hospital was already crowded with patients battling these conditions, although we weren't entirely sure why. Higher temperatures may allow certain bacteria to thrive, and people are more likely to be outside, walking around with open-toed shoes, exposed to all sorts of pathogens.

Tom patted me on the back. "The real work," he said, "is about to begin."

PART 3

The Volunteers

CHAPTER 10

Ruth

RUTH KNEW why the men had gathered outside of her home, and she knew what to do. The order to confine Hungarian Jews to ghettos had been signed several weeks earlier, on April 7, 1944, and Ruth figured she had only a few moments left. She grabbed the pair of ruby-red shoes her father had just given her—an unexpected sixteenth birthday present—ran upstairs to the attic, and placed them under an old mattress. Then she closed her eyes, committing the location to memory, and raced down to the living room, where the rest of her family had gathered. Ruth's younger brother put an arm around her shoulder and gently squeezed. They looked up at their mother, who was whispering to herself. A moment later, there was a knock at the door.

The first anti-Semitic law had gone into effect in Hungary six years earlier, on May 28, 1938, when a 20 percent ceiling was placed on the proportion of Jews who could work in finance, commerce, and most other businesses employing more than ten people. Ruth's father worked at a small bank, and despite the rising tide of anti-Semitism across Europe, business mostly went along as usual. But the following year, as the ultraright consolidated power in Hungary, another law was passed that dropped the employment cap to 6 percent, and Ruth's father lost his job.

She was eleven when he was fired, and didn't realize what was happening. Few did.

Ruth attended school and played with friends like she always had. Things were mostly the same, she told me—or perhaps just slightly different. At first, Ruth was glad to have her father at home more, and didn't recognize the increasing stress her family was under. Then, in June 1941, he was conscripted into the war effort. He wasn't allowed to handle a gun; instead, he spent his days repairing roads and stocking storerooms. He spent months saving his meager salary to buy Ruth that pair of shoes for her birthday. By then, Hungary had entered into an uneasy alliance with the Nazis, and the country was no longer a place for displaced European Jews to seek refuge. In the spring of 1944 Hitler had become convinced that Hungary, an independent member of the Axis powers, was trying to dissolve the partnership, and he pounced. The Nazis expeditiously moved east to occupy the country. That's when the roundups began.

Ruth's friend Edith told her that some families were being taken away, but she didn't know where or for how long. The two girls, friends since they were toddlers, decided to hide their prized possessions—shoes, books, a few dresses—in their respective attics and promised to look after the other's belongings. With the gendarmes gathering outside of her home, Ruth knew her family was next. The knock preceded a pounding on the door. A moment later, it flung open. Ruth's family would soon board a train to the Auschwitz concentration camp in Poland.

Ruth rarely spoke of what happened after that train ride, and when she did, it was never in detail. Almost half of the Jews killed at Auschwitz were Hungarians, who were gassed within a ten-week period during the summer of 1944—including Ruth's parents and her friend Edith. In fewer than four months, most of Hungary would become *Judenrein*: free of Jews.

After Auschwitz was liberated in spring 1945, Ruth did the only thing she knew to do: she went back to Hungary and to her family's small home near the Miskolc ghetto. She knew the shoes would be gone, but she felt compelled to check the attic. The house was empty, of course, all of her possessions having been seized and destroyed. She never found the final gift from her father.

A few years later, Ruth moved to Paris and then to Lisbon, Portugal. She married an academic—a professor of romance languages—and had a daughter. Over the next five years, she would have three more. She spent the postwar years as a seamstress and, later, as a schoolteacher. She had very few indulgences, she said, but one of them was shoes. She delighted in shopping for them and always made sure her four children had fashionable footwear. In the late 1960s, Ruth's husband was offered a job at a university in New York, and they moved to Brooklyn, where she's lived ever since.

A week before I met Ruth, she had gone shopping with her daughter, Anne, for a pair of bedroom slippers. During the outing, both women had been bitten by a swarm of mosquitos. A day later, Ruth noticed some swelling and redness around the bite on her left ankle. The next morning, the discoloration had snaked down to her foot and up toward her shin. She paid a visit to her primary care doctor, who suspected a bacterial infection, and prescribed an antibiotic called Bactrim. It was a sulfa drug, a distant relative of sulfanilamide, and it was supposed to treat her serpentine skin infection.

Over the next few days, the pain and swelling worsened, and despite her initial misgivings, Ruth went to the NewYork-Presbyterian Hospital emergency room. After her vital signs were taken, Ruth was placed on a stretcher and examined by a group of medical students, residents, and nurses, until the attending physician appeared and diagnosed her with a drug-resistant skin infection called cellulitis. The doctor suspected a MRSA infection and admitted her to the hospital for treatment with an intravenous antibiotic. And then Ruth waited.

She spent the next twenty-three hours on a stretcher in the hallway of the bustling emergency room hoping for a room to open up. At a quarter after four in the morning, she was finally wheeled up to the fifth floor. Ruth would share her room with a twenty-seven-year-old intravenous drug user who had contracted metapneumovirus, a form of viral pneumonia, and was coughing behind a surgical mask. "I'm Dr. McCarthy,"

I said when I walked into Ruth's room a few hours later, "and I'm running a clinical trial." Her bed was next to a large window facing the East River, and sunlight was starting to filter into the dimly lit room. Ruth was surrounded by family: her daughter, Anne; Anne's husband, Michael; Michael's sister; and a throng of grandchildren and distant cousins. Plush throw pillows, a homemade quilt, and a half dozen stuffed animals were scattered across the bed. Ruth was holding a pen in one hand and a breakfast menu in the other. I had seen her name on a list of patients in the emergency room with skin infections, and I thought she might be the first to qualify for my trial. I knew very little about her, but that would soon change. "Should I come back?" I asked. "I don't want to interrupt."

Ruth's chart noted that she was a Holocaust survivor, but there were no further details. I thought about excluding her from my study because of this—I felt uncomfortable asking anything of someone who had been through so much—but the protocol didn't include provisions that would make this possible. I was evaluating consecutive patients for strict inclusion criteria, and she met them. I couldn't pick and choose whom to study; I had to approach everyone. That was crucial for the integrity of the research. Cherry-picking the most treatable cases would compromise the data. Ruth shook her head. "Come in."

The children cleared a path as I approached with my stack of consent forms. For the first part of the trial, I was observing patients to understand how physicians normally treat drug-resistant infections such as MRSA. This information would give us a baseline—a starting point—and an idea of how we could improve things. I suspected that patients with infections were hospitalized longer than necessary due to provider uncertainty: physicians were afraid of discharging their patients prematurely and erred on the side of keeping them under observation. Uncertainty caused delay, and delay led to complications. I wanted to know more about those complications so that I could figure out how to prevent them.

I explained the details of the trial and asked Ruth if she might want to participate. I would check on her two days from now, two weeks from now, and six weeks from now to ensure that the infection had resolved. Ruth looked at her daughter and then nodded at me. "Okay," she whispered. Soon a grin emerged. Several teeth were missing, but Ruth's smile was still radiant. She put down the menu and I handed her a consent form. "Do you mind if I examine you?" I asked.

"She's having trouble swallowing," Anne said. "Is that something you can fix?" Ruth's daughter was wearing a blue-and-yellow floral-print dress and there was a pillbox hat in her lap. "It's even hard for her to take small sips."

"Is that right?" I asked. The words sounded awkward coming out of my mouth. I tried to push away the thoughts of what Ruth had been through in Poland, but I couldn't. I felt like I was handling a delicate artifact. It was objectifying to view her in this way, but that's where my mind went. It was impossible to ignore the historical significance of Ruth's life. As I stared into her rheumy grey eyes and prepared to begin my assessment, a swell of emotion bubbled up that left me unable to speak. I forced the words out. "Are you . . . okay?" It was a question I was also asking myself.

Ruth nodded. The swallowing problem may have explained why Bactrim, the oral antibiotic, had been ineffective. Perhaps she wasn't absorbing it properly. I placed my stethoscope on Ruth's tiny chest and listened to her heartbeat. Systole and diastole. The heart contracts, expelling freshly oxygenated blood throughout the body's circulatory system; then the cardiac muscle relaxes so that the organ can expand and refill with blood. Back and forth it goes. "A few deep breaths," I said, shifting my focus to her lungs. I could feel sweat gathering under my arms. "Deep breath," I said to both of us.

"Do you think she needs a feeding tube?" Anne asked. There was a note of tension in her voice, as though she'd asked that question before and had not received a satisfactory answer. I wanted to help, but I occupied a special place in Ruth's care: I was an investigator, not a caregiver, and it wasn't my place to intervene or arrange for procedures. I was there to talk about infections.

"I'm not sure," I said. "Let me talk to the hospitalist caring for your mother." In the old days, primary care doctors cared for their patients when they were hospitalized. But in the 1990s the pace of medicine changed, and most physicians weren't able to manage an office *and* care for hospitalized patients. A specialty was born—hospital medicine—staffed by hospitalists, doctors who specialize in inpatient care. It has since become the fastest-growing subspecialty in the history of medicine. A hospitalist would sort out Ruth's swallowing problem, not me.

"Our rabbi," Anne continued, "he thinks she should get the feeding tube. Do you mind if we talk outside?" I looked at Ruth, and she indicated that it was okay. Anne put her arm on my elbow and ushered me out of the room. We walked to an alcove near the nursing station, where she let out a small sigh. Anne had blond hair and deep lines etched on her forehead. She held the pillbox hat in her left hand. "Is my mother going to be okay?" she asked. Nurses and doctors hurried past us as I studied her face. She had been at her mom's side all night. "Is she going to survive this?"

"Yes," I said, with a bit more certainty than was warranted. "She'll get through this. And I'll find out about the feeding tube."

"Thank you."

"I know it's horrible being down in the ER," I said. "I'm sorry for that."

Many children of Holocaust survivors report recurrent dreams in which they are chased, persecuted, and tortured, as if they have absorbed their parents' harrowing pasts. One controversial theory suggests that the DNA of survivors was altered by their experience in the concentration camps, allowing a transgenerational transmission of trauma. (A number of prominent scientists say that this simply cannot be true.)

"I'm sure you saw her chart," Anne said.

"I did."

"There's a lot . . . there's a lot about my mom that's not in there. Not in the chart."

I nodded. "I'm sure. I wouldn't know where to—"

"They experimented on her." She looked away briefly, as if she had said something improper. "She won't talk about it, but I thought you should know." Ruth had been a *Versuchsperson*. That wouldn't rule her out of my study, but maybe it should have.

"I'm not sure this trial is right for her," I said.

Anne shook her head and smiled. "It's okay. I'm sure she'll do it. Just ask about the feeding tube."

"Of course."

I returned to the room, and Ruth waved me in. I pulled out a consent form and took a seat in a red plastic chair next to her bed. As she and I chatted, I began to understand the contours of her remarkable life, from Poland to Brooklyn, and everything in between. When I mentioned the ruby-red rash on her leg, it reminded her of the red shoes her father had given her. She spoke about the day her family was taken away and explained what happened in the days following the roundup. But then she stopped. Ruth quietly asked that we talk about something else. Eventually I finished my examination. No scars were visible. They were all on the inside.

Obtaining informed consent is rarely straightforward. It typically involves a nuanced conversation with an exhausted, vulnerable patient, and (hopefully) an extended question-and-answer segment: Is the treatment safe? Who is funding the trial? Why is it being performed? But some patients fear they'll receive substandard care if they refuse to participate, or if they ask too many questions. Silence is a red flag. Conversely, if someone appears *too* eager, I offer to come back later to discuss the risks and benefits in greater detail. Ruth quietly asked a few questions and, with her daughter's blessing, agreed to participate. I mentioned that we would modestly compensate the participants (with a $200 debit card), and her son-in-law suggested that she might use it to buy a new pair of slippers.

When I returned to see Ruth the next morning, though, there was a problem. In the middle of the night, she had become confused and tried to crawl out of bed. (Delirium is common in elderly hospitalized patients.) On the way to the toilet, she stumbled and landed awkwardly on her shoulder. An X-ray at three in the morning ruled out a fracture, but it revealed a small nodule on her lung. The family was trying to figure out what to do. So were the doctors. There was no update on the feeding tube.

This was one of the many complications that I had feared. The hospital is a dangerous place, especially for an older person. Accidents happen, and diagnostic tests often lead to more tests, which may or may not prove useful. This was one of the reasons I was performing my study: to understand what happens when a patient with a drug-resistant infection enters a hospital. Once that baseline was established, I would start administering dalba. For Ruth, standard medical care meant an endless stretch in the emergency room, followed by an accidental fall and a battery of unexpected tests. It meant more questions than answers.

I knew we could do better.

CHAPTER 11

George

A FEW WEEKS before Ruth's family heard that terrible knock on the door in 1944, a young man from Missouri boarded a military plane bound for New Guinea. George Hermann had enlisted on a whim, hoping the military would provide a change of scenery and, as the saying goes, a chance to be a part of something larger than himself. The small-town boy felt very small indeed as he flew west over the Pacific and eventually approached the massive island—one of the most isolated in the world. As the plane descended upon the lava-encrusted isle, he saw tiny villages ravaged by bombs and torched by flames. A few minutes later, he was on the ground, ready for war, trading the tranquil life he'd known in the Ozarks for one of steel helmets, air raids, and tommy guns. There had been no second thoughts or fraught good-byes when he'd left home in the early spring. "We had a job to do," he told me, "and everyone knew that."

George was stationed near the equator, surrounded by a thick jungle, a malaria-infested swamp, and a vista of cascading mountains. After he was settled, his gaze turned to the sky, which was filled with the screech and snarl of airplanes: P-40 fighters, fabric-covered biplanes, and Japanese dive-bombers. During his eighteen-month tour of duty, he would serve as an air observer, seated behind a pilot in a Piper Cub propeller plane, day after day, searching for targets. When one was identified, he radioed down to men on the ground and told them where to aim. If they missed, he'd tell them by how much, and coordinates were adjusted.

The Japanese were more familiar with the art of jungle warfare and had a tactical advantage on the ground, while the Americans were vastly superior in the air. But in the trenches the enemies had a common adversary: tropical infection. Diseases such as yellow fever and mosquito-borne encephalitis had spread through both camps, incapacitating five times as many soldiers as actual combat. American medics were overwhelmed with men suffering from diarrheal illness, fungal infections, and unremitting fevers. George had never seen anything like it, and did his best to stay healthy. But remedies were hard to come by. He watched with dismay as dysentery and ulcerative skin infections spread through the barracks, and he learned how to protect himself during downpours. He became a "neat freak," he said, conscientious of hygiene and avoiding person-to-person spread of germs. That grew more difficult after the war, when he moved to a place that was even dirtier and in which people lived in even closer quarters than in the barracks in New Guinea: New York.

George had been surprised when, nearly seventy-five years later, his primary care doctor on Long Island diagnosed him with a skin infection and attributed it to MRSA. "I don't know how I got it," he said, "or who gave it to me." George was prescribed a ten-day course of clindamycin, an oral antibiotic to treat MRSA cellulitis on his right forearm. Several days into the course, however, he developed explosive diarrhea. He was moving his painful bowels every hour, on the hour, and was unable to eat or drink. He felt feverish, he saw flecks of blood in his stool, and his heart started to race. When George stood up from the toilet one evening, he nearly passed out.

He called his physician for advice. The doctor suspected the diarrhea was from a bacterial superinfection called *Clostridium difficile* and told him to stop the antibiotic, despite the fact that the skin infection was still spreading. Clindamycin can be remarkably effective—it has been on the market for fifty years and is on the WHO's list of essential medicines—but it can also cause unfortunate side effects, including *C. diff* diarrhea. While the drug was destroying some of the bacteria on George's skin, it was also wiping out a large swath of the good bacteria living in his colon. That cleared the way for *Clostridium difficile* to pro-liferate, producing a toxin that induced massive diarrhea. The complex interplay between body and bacteria, known as the microbiome, had been disrupted, and it nearly killed George.

The doctor wisely instructed his patient to go to the nearest emer-gency room for aggressive rehydration. George was in a taxi when he received the call, and instructed the driver to drop him at NewYork-Presbyterian. Within thirty minutes of his arriving, the staff started an intravenous antibiotic called vancomycin, the same one Ruth was re-ceiving, as well as an oral drug called metronidazole to treat the *C. diff.* Then the ninety-six-year-old waited.

Nineteen hours later, George was notified that a bed had become available, and he was wheeled from the ER to a room on the fifth floor, just across the hall from Ruth. Not long after he arrived, I introduced myself and asked George if he'd like to enroll in my trial. His skin was weathered, his mucous membranes were dry, and there was a large red patch on his forearm from MRSA. Just below the rash, a bag of saline was dripping into an IV that tunneled into his right hand.

"What can you do about the *C. diff*?" he asked as I stepped into his room and pulled out the consent form. "Maybe a probiotic?" He was bald, with sunspots on his scalp, and had bright blue eyes and a large scar on his left cheek—a souvenir from the war. "Think that'll help?" Probiotics are marketed as the "good bacteria" that keep your gut healthy and potentially help with digestion, depression, and heart health.

Most clinical trials had shown minimal benefit to probiotics for someone like George—it's hard to figure out which good bacteria are missing—but some patients swore by them. "Maybe," I said. "We're still waiting for the *C. diff* test results. I'm here because you may qualify for a clinical trial." I sat down and explained the study as I had with Ruth. "Feel free to say no," I added.

He read the consent form and then studied my face. "Why wouldn't someone do this?" he asked. His voice had a gravelly quality to it, as though his vocal cords had been plucked by a calloused hand. "The trial. I mean . . . why not?"

One of the first things Tom Walsh taught me about running a clinical trial is the rule of halves: half of all patients who meet inclusion criteria will also meet exclusion criteria. They'll have an allergy or a preexisting medical condition that makes them ineligible. Of the remaining patients who meet inclusion criteria, half will simply decline to participate. So for every four patients who qualify for a study, perhaps one will enroll. "Some people don't want the hassle," I said to George. "Or they don't want to be a guinea pig."

It was that phrase—*guinea pig*—that sparked his memory of World War II and New Guinea and launched us into a conversation about dive-bombers and tommy guns. It was the same reaction that I had elicited from Ruth when I said "ruby-red." As George spoke, I tried to keep up, jotting down whatever I could about his time in the war and the lifetime he'd spent in New York. He was a Yankees season ticket holder and was once a scratch golfer; he liked Tiger Woods and owned two parakeets: Frick and Frack. Perhaps the most striking thing about the man seated in front of me was his robust health. At ninety-six, George had just two medical conditions—high blood pressure and atrial fibrillation—and took only three medications. "You qualify," I said, "if you're interested."

He looked over the paperwork and nodded. "Sure. Why not?"

George told me that his first meaningful interaction with a doctor occurred shortly after he joined the military, and he's trusted them ever since. During the standard physical that all recruits endure, George learned that he had a heart murmur. "I was kinda scared at first," he recalled, "but the medic was really nice about it. Said there wasn't anything to do. And he was right; all these years, it's never been a problem." His faith in medicine was forged in that moment, he said, when a physician shook his hand, looked into his eyes, and offered him reassurance.

This happened to George around the same time that a military physician was looking into the eyes of a black recruit in Alabama, telling him that there was nothing to do about his syphilis. Or, perhaps, that doctor said nothing at all. "Here you go," George said, handing the signed consent form back to me. "Happy to help, Doc."

George and Ruth had spent the war on opposite sides of the world in vastly different circumstances, but now they were just across the hall from each other, confronting the same skin infection. And now both were enrolled in my trial.

CHAPTER 12

Mississippi Mud

RUTH AND GEORGE were both receiving vancomycin, a cheap antibiotic that's been around for more than sixty years. It's one of the most commonly used drugs at my hospital—we order it when a patient has an infection that's spreading and we don't quite know why—but in recent years, its power has waned. Bacteria are scavenging for genes that weaken or degrade vancomycin, and the wonder drug doesn't work like it used to. That's where Allergan comes in. We need something new, and if dalba can replace vancomycin, it will become the kind of blockbuster product that Anthony Fauci speaks of: the kind that recoups the losses associated with all of the other failed drugs.

Vancomycin once held this distinction. A few years after penicillin hit the market in the 1940s, doctors noticed that bacteria were developing ways to avoid it. When penicillin binds to a chemical in the bacterial cell wall, it compromises the entire structure—imagine a tiny grenade inserted into a Jenga puzzle—and it prevents the infection from spreading. But after even brief exposure to penicillin, bacteria can subtly change shape and the drug no longer works. The Jenga puzzle evolves from wood to brick-and-mortar, and the grenade no longer fits.

Just a few years after penicillin hit the market, it became obvious to physicians that they would need something else to treat patients with infections. But where to find it? There simply wasn't time to wait for another Alexander Fleming or another serendipitous discovery in a laboratory, so pharmaceutical companies sent teams of researchers out to scavenge the globe, searching for something—anything—to treat bacteria that had become resistant to penicillin.

In 1952 a missionary in Borneo sent a sample of dirt to his friend E. C. Kornfield, an organic chemist at the Eli Lilly company. Buried within the sample was an organism called *Streptomyces orientalis* that made a substance, later called compound 05865, that could kill penicillin-resistant bacteria. Compound 05865 was extracted from the dirt sample using a technique called chromatography, which dissolves molecules based on their size, acidity, and electrical charge. The purification step yielded a new drug that the chemists at Eli Lilly called "Mississippi Mud" because of its brown color. They evaluated it in test tubes and then briefly in animals, and determined that the mud should be tested in humans. A bit of chemical tinkering removed some of the impurities that caused the muddiness, and the new translucent drug was rechristened as vancomycin, its name derived from the word *vanquish*. It took just six years for the dirt from Borneo to receive FDA approval.

Vancomycin was initially reserved for patients with severe penicillin-resistant infections, but it soon found its way into everyday use. It was simply better than its competitors, and doctors weren't willing to give their patients inferior treatment. But the drug wasn't without side effects. Vancomycin can cause kidney problems and hearing loss, and some people develop an allergic reaction known as red man syndrome, which usually begins a few minutes after administration and may cause a rash that affects the face, neck, and chest. Today we closely monitor patients receiving vancomycin, taking frequent blood samples to ensure that they don't receive too much or too little.

Blood tests are a hassle, especially for patients who may require months of antibiotic treatment, and Allergan has been betting that dalba can steal a slice of vancomycin's market share. There are no blood draws or repeated injections. Patients get one infusion and walk away. It's revolutionary, in theory, until we factor in cost. Vancomycin is generic, and its wholesale price is about $40. Dalba costs several thousand dollars per dose, and it's unclear how our cash-strapped health care system can pay for it. It's not all that surprising that Allergan is charging a hefty fee for a new drug. But how much is too much?

NOT LONG after I enrolled Ruth and George in my trial, I wandered down to the ER to meet a young man named Erwin Davis. He was a fourth-year medical student from Nebraska who had come to Manhattan to audition for a residency training spot at a neighboring hospital. This is common for the more selective specialties—dermatology, ophthalmology, radiation oncology—and it was necessary for Erwin's field, neurosurgery. He had bright green eyes and a bowl cut, and when I approached his stretcher, he was cozied up next to a woman wearing heavy eyeliner. "I'm Dr. McCarthy," I said, "and I'm running a clinical trial."

Erwin's face lit up. "Nice," he said. The woman pulled away and took out her phone. I retrieved a consent form and considered my words. "It's a cellulitis study," I said. "I understand you have a skin infection?" Looking into his eyes, I recalled how I had felt at his age, on the cusp of leaving medical school and becoming a doctor. There was a mix of terror and excitement as medical graduation approached, knowing that people would soon put their lives in my sweaty hands. I still felt that way. "I hear you're a medical student."

The young man pulled up his shirt, and his friend rolled her eyes. Beneath the makeup, she appeared quite young—possibly a teenager. "Right here," he said, pointing to his right nipple. "It's pretty bad." The area was swollen and inflamed—his doctors would call it "angry"—and it appeared twice its normal size. I drew a curtain and peered at his chest. "It looks painful," I said. "What happened?"

Erwin smiled and looked at his companion. "Tequila shots." The young woman rolled her eyes again as she stared into her phone. "We got a little crazy last night."

"You're an idiot," she said, and they both laughed.

The boozy details were actually crucial for my study. "We're gathering information on patients with skin infections," I said, "people who require hospitalization. But there are a number of exclusion criteria."

"Like what?"

"Bite wounds are excluded. Human and animal." The mouth contains a variety of bacteria that might not respond to antibiotics such as vancomycin or dalba. (Some doctors fear cat bites more than dog bites because their teeth are sharper and more likely to puncture bone.) Erwin raised an eyebrow. "Was there biting? Certainly a lot of licking. But I'm sure—"

His partner put down the phone and sighed. "There was no biting, sir."

I nodded and handed Erwin the consent form. After a minute or two, he handed the unsigned document back to me. "Let's role-play," he said. "I'll be the patient."

"You *are* the patient," I reminded him.

"Loves to role-play," the woman said softly.

Had I acted this way when I was a medical student? I imagined how different Erwin would seem a year from now, after a year of hundred-hour workweeks, devastating feedback from superiors, and a stream of sleepless nights. Intern year had permanently changed me, and it would change Erwin, too.

"Have you achieved equipoise?" he asked. Erwin was in a playful mood, perhaps delighted to have the roles reversed, to be a student questioning the professor. I could tell he was proud of himself for using the word. "It's of utmost importance," he said with faux seriousness. He was referring to a state of uncertainty about the relative benefits of various treatments in a clinical trial. If equipoise exists, no enrollee in a randomized study is knowingly given inferior treatment. "It's a must," he added.

"I see you're familiar with research ethics," I said.

He looked at his companion and then back to me. "A little bit." As we spoke, Mississippi Mud slowly dripped into his IV. "If there's no equipoise," Erwin went on, "you shouldn't do a trial. It's not fair to the patient." He was trying to impress the woman, but it didn't appear to be working.

I started to imagine Erwin out of his hospital gown and in a long white coat, a buoyant physician tossing off medical facts on rounds. "I actually teach ethics," I said, "to our first-year medical students."

"Oh, yeah?"

"Do you know the problem with equipoise?" Erwin shook his head. "Let me rephrase that. Do you know its limitations?"

"I don't."

"Well," I said, trying to decide how much time I wanted to devote to an impromptu didactic session, "who gets to define it? A group of experts? Who selects them? And how do you define uncertainty?"

"Huh."

"Seems obvious, but it's not. How many experts need to disagree?"

"Two?" he asked.

"It's a rhetorical question." I returned to the consent form. "There's uncertainty with this trial, but I can't say whether we've achieved equipoise."

"Gotcha." The other problem—one that I chose not to discuss with Erwin—is that equipoise promotes the early termination of trials. If a data monitoring board believes that treatment A is better than treatment B in the midst of a study, the board can shut it down. And that can compromise results. I had once been drawn to equipoise, but I have come to know its distinct drawbacks. "Let me check out the form again," he said. "It's a drug study?"

As the words came out, I noticed Jackson across the hallway. He was in a wheelchair now, accompanied by a woman and two small children, and a nurse was about to take his vital signs. He removed his puffy black jacket and disconnected a small metal oxygen tank at his side as I handed Erwin the consent form. Colistin had halted Jackson's initial infection, but it would never cure him. The man looked like he'd aged a decade since I'd last seen him. His eyes were tired, liked mine, and he winked when he saw me. I waved and mouthed to him, "One minute."

Erwin and I reviewed the consent form together, line by line, until we reached the last page. "For your participation," I said, "we'll give you a two-hundred-dollar debit card at the conclusion of the study."

"For what?"

"As a thank-you."

His green eyes glimmered. "Yes, please!"

The languorous pace of research was something I often struggled with, moving along in unpredictable fits and starts, but now it felt like genuine momentum was building. Three patients in a row had opted to enroll. Erwin signed the paper and handed it back to me. Then he asked his companion how they should spend the money. "This is great," he said as I walked away. She agreed.

I tucked Erwin's form into my white coat and wandered over to Jackson, grabbing a protective gown and disposable gloves along the way.

CHAPTER 13

Soren

DESPITE ERWIN'S ENTHUSIASM, something about the exchange rubbed me the wrong way. Perhaps it was that he hadn't taken the consenting process all that seriously, or that money may have influenced his decision to participate. As I hewed to the script, explaining the history of dalba and the expectations of the trial, his face had drained of expression, like a student bored at a lecture. It was an unfortunate reality—some patients were going to enroll for the cash—but I still had to ensure that a signature represented informed consent. Money obscured that.

Financial incentives used to be controversial—my medical school classmates fiercely debated the concept—but now it's a routine aspect of recruitment. So much so that when I negotiate with pharmaceutical companies, one of the main sticking points is how much to pay volunteers. More than half of all research subjects are paid to participate, and in my experience, it's the rule, not the exception. Some argue that the financial transaction renders the term *volunteer* obsolete, but if so, what should we call the folks who agree to be studied?

Filling out questionnaires and returning for follow-up visits is a hassle—it can take an hour just to find parking near my hospital—and people *should* be compensated for their time. But it opens the door to coercion, especially for patients from marginalized groups. Researchers in the United States have consistently found that moderate incentive payments are effective at improving recruitment without being undue or unjust inducements. For the most part, I agree.

A day after I enrolled Erwin, I met Soren Gillickson. Three years earlier, the thirty-one-year-old computer programmer had been in a car accident on East Fifth-Seventh Street and fractured his femur. He'd been rushed to a nearby emergency room, where a team of orthopedic surgeons went about the task of repairing his left leg. A blood clot had lodged in major veins, causing a condition called phlegmasia cerulea dolens (PCD); his leg swelled and turned blue, and doctors worried that he might lose the limb. The six-hour procedure, which ultimately required the assistance of vascular surgeons, was a success, and Soren's doctors anticipated a speedy recovery. But complications began shortly after he left the operating room.

Soren told his nurse that he was experiencing excruciating pain at the site of the incision. The nurse alerted the covering doctor—a second-year physician with marginal knowledge of Soren's condition—who prescribed a dose of OxyContin, a popular opiate. The pain subsided, and Soren slept through the remainder of night. But the following morning, the pain returned, and he asked for (and received) a dose of Oxy. A day later, a physical therapist arrived to put him through a workout. He wasn't experiencing discomfort when she arrived, but his surgeon anticipated that it would be a painful experience, so he was given another dose of Oxy prior to the therapy session.

It worked. Soren was able to make his way around the hospital unit without pain, and the following day, he could walk up a flight of stairs, albeit with a double dose of Oxy and a single dose of an even more powerful painkiller, Dilaudid. When it was time for discharge, Soren was given a thirty-day prescription and sent on his way. When he ran out a month later, he returned to his doctor, who denied the refill request. The leg had healed nicely, and there was no reason to extend the prescription. But Soren had become physiologically dependent on Oxy; when he ran out, his heart began to race, and he developed intractable diarrhea.

He was withdrawing from two opiates, paralyzed in a cold sweat, unable to eat. So Soren turned to the street.

It took him less than an hour to find what he was looking for, which set in motion of a cascade of events that has been playing out across the country. Since 2000, two hundred thousand Americans have died from overdoses related to prescription opiates such as OxyContin, and more than three-quarters of people who try heroin, as Soren ultimately did, started with prescription painkillers. His path to addiction began in a hospital.

When he stumbled into my ER, Soren was a full-blown addict. The first thing I noticed was that his right hand trembled when he ran his fingers through his thick, dark hair. He had a skin infection covering his left forearm, and his chiffon-yellow eyes were barely open. His face was gaunt, and a webbing of wine-colored blood vessels was visible just below his thin, pale skin. Soren looked like a vampire with a sunburn.

Soon I was examining a large rash just below his elbow. "Hurts like hell," he said as I measured it. Soren was not receiving vancomycin like the other patients; instead, he was getting a combination antibiotic called ampicillin-sulbactam, also known by its brand name, Unasyn. Ampicillin is a variant of penicillin that was developed in the early 1960s; it was later paired with sulbactam, a bacterial enzyme inhibitor, to boost its strength.

"It hurts," Soren said. "Damn, it hurts." His raspy voice was difficult to hear amidst the riotous din of the emergency room. He squinted as I pushed on the infection, trying to express pus. "Aaaah," he said. "Stop, man."

"I'm sorry." I pulled away my hand. "How long have you had this infection?"

"Couple days, I guess. Maybe a week?"

Unasyn is an excellent drug, but it doesn't kill some of the more aggressive superbugs such as MRSA. I wondered if I had an ethical obligation to tell him that or to intervene at all. I wasn't his doctor, I was a researcher, and there is a firewall between patient care and experimentation. I was documenting what happened to patients—good or bad—

but I didn't want to ignore his pain or the rapidly spreading rash that was about to encase his lower arm. "I'm running a trial," I said, "and you might qualify. I need to talk with your doctor, too."

"I'm game," he said. "Just stop pushing on my arm." Soren reached into a black backpack near his feet to grab a pen. "Where do I sign?" His eagerness gave me pause. Like Erwin, he was a bit too willing. He didn't know anything about the study. I suspected he might be agreeing just so I'd leave him alone. "We should go over a few things. We can do it now or later."

I offered to come back at a different time, but he waved away the suggestion. "Do your thing."

With every movement, his hands were shaking. Soren's voice was calm, but he was in the throes of withdrawal. I wasn't sure he was capable of giving informed consent. "Do you feel okay?" I asked.

"Been better." He took a deep breath and tried to open his eyes. "Can you get me something for the pain?" His fingernails were caked with dirt, and his pants had a large tear near the inseam. Soren's breath had an ammoniac whiff, a possible sign of kidney dysfunction, and his dilated pupils darted between mine as we spoke. Examining him, I felt a mix of emotions—sympathy, frustration, regret—but mostly sadness. Soren was worn down, exhausted, beholden to a chemical that wouldn't let him go. "We'll get you through this," I said. "Let me talk to your doctor about the pain and the antibiotic."

"Thanks, man."

"The thing with this trial," I said, "is that you have to follow up with us. Two weeks from now, six weeks from now—it's pretty involved."

"That's fine."

"How do you spend your days?" I asked. "Student? Do you work?"

"Used to."

I tried to imagine this other Soren, well rested and happy, sitting at a desk and typing on a computer. "What about now?" I asked awkwardly. "What do you do?"

"Nothing."

"Before I go over the consent form," I said, pointing at the document, "I have to ask a few screening questions."

"Fine."

"Do you smoke?"

"Nope."

"Drink?"

"Negative."

"Allergies?" Soren raised his thin arm and showed me the word on his hospital-issued wristband: *Sulfa*.

"What happens if you get sulfa drugs?"

"My skin falls off," he said, "or tries to."

On occasion, sulfa drugs cause Stevens-Johnson syndrome, a devastating side effect where skin blisters and melts away. It was one of the last things the Massengill company had noticed about its drug. Patients start out with vague, flu-like symptoms and end up in the burn unit. "No sulfa drugs for you," I said. "A few more questions; we're almost done."

"Take your time."

"Do you, ah, use drugs?"

"Yep."

"Illicit drugs?"

"Like what?"

"Do you use IV drugs?"

A smile bloomed across his face. "When I can."

"Recently?"

He looked at his watch. "What's today?"

"Wednesday."

"Well, then, yes."

"Give me one minute," I said. I stepped away from his stretcher and pulled up his chart. When Soren had entered the emergency room, the triage nurse had taken a set of vitals and found that he had a fever of 101.7 degrees Fahrenheit. A fever in the setting of intravenous drug use can represent a life-threatening infection of the bloodstream or the heart and would require a different treatment the type of skin infection

I was studying. "I'm sorry," I said a moment later, "but you meet one of the exclusion criteria."

"Okay."

"It means you can't enroll in the trial."

He shook his head. "Bummer."

His infection was more serious than he realized. I didn't want the medical system to fail this man again. Soren looked down at his elbow and then up at me. "Okay if I rub it?" he asked. "It's really starting to hurt."

"Yeah," I said, letting out a small sigh. I put a hand on Soren's left shoulder and moved my palm in a circular motion, still thinking about what he was like before addiction took hold, happily typing away in a bustling office. "Let me go find your doctor. We need to try something else." Two hours later, Soren was back in the operating room, chemically sedated as a team of surgeons carefully washed out his infected elbow.

CHAPTER 14

Duty

"I KNEW as soon as the first one hit," Donny Alexakis said. He was supine on a stretcher in the emergency room, vigorously scratching his right forearm as he spoke. "Couldn't have been an accident. Not on a clear day, you know?" Donny was in his late fifties and had tiny red spider veins known as telangiectasias all over his face. He was wearing a grey New York Giants T-shirt and white sweatpants when we met shortly after my encounter with Soren. "I immediately started crying," he went on, relaying how the nightmare of 9/11 had played out for him. "And then," he said, sitting upright and leaning toward me, "I started running around the house like a madman, grabbing my gear, grabbing my boots, anything I could find." The next morning, he reported for duty.

Donny had spent twenty-two years as a New York City firefighter, and when the towers were struck on that Tuesday morning in September, he was sitting on his deck in the mountains of Kentucky, enjoying the first few months of retirement. "They called in everybody," he said, "and told us to report to our firehouses." He was assigned to a command center on Vesey Street, not far from the smoldering rubble, where he began work as a liaison for the deputy fire chief. Yanked out of retirement, Donny was soon working twelve-hour shifts, ensuring that the NYFD coordinated with the NYPD. "While I was down there," he said,

"I breathed in a lot of toxic stuff. Didn't know it at the time, but they said I was breathing benzene."

We've since learned a lot about the air Donny was breathing: there were roughly seventy carcinogens in the smoke that emanated from the fire and debris, and now, more than fifteen years later, the first responders are developing all kinds of malignancies, including brain, bladder, and breast cancer. Those fumes contained heavy metals such as lead and mercury along with asbestos fibers and volatile compounds like benzene, a chemical solvent I'd used in the laboratory to mutate DNA. In this experiment, Donny's body was the test tube. "I had no idea what was happening," he said as he continued to scratch his arm. "None of us did."

Donny was one of more than fifty thousand rescue workers potentially exposed to poisonous chemicals in the weeks that followed. In addition to cancer, many developed chronic, progressive, unrelenting physical and psychological syndromes that are largely untreatable. One pair of first responders, a father and son, died of ground-zero-related cancers less than a year apart.

In 2010 the Zadroga Act—named after James Zadroga, the first police officer whose death was attributed to toxic chemicals after the attack—established the World Trade Center Health Program to monitor and treat the surviving first responders. It was through that program that Donny discovered he was in trouble. "I got a call, and my world changed," he said. "I don't remember the details, but the gist of it was that I had bad blood." Those two words sent my mind ricocheting to a different place, far from our emergency room or the smoldering rubble in Lower Manhattan to a small town in Alabama. "They brought me in for more tests—all kinds of stuff. They told me my white blood cell count was off the charts." A bone marrow biopsy confirmed the diagnosis: leukemia.

"Looking back," Donny reflected, "what they shoulda done was call in the military. Bring in guys with hazmat suits. But that woulda panicked the country. That would've started riots. They couldn't do that, but maybe that's . . ." His voice trailed off. "My body's so weak," he said, pointing to his finger. "A friggin' paper cut lands me in the hospital." He stopped scratching and showed me his throbbing red index finger, which was twice the size of the others. It looked like a sausage about to burst. I gently squeezed it, trying to express pus, just like I had with Soren. Something so small—a simple paper cut!—could wreak such havoc in a patient with leukemia. Chemotherapy had ravaged his immune system, and a little nick could serve as a portal of entry for bacteria. "I used to be a Greek god," he said with a wise-guy smile, "now I'm just a goddamn Greek."

I pulled out my stethoscope, repeating a phrase that I said to patients whether I believed it or not. "We'll get you through this." A nurse and an orderly gadflied nearby. "One sec," I said. "I just need one more minute."

"I poured peroxide on it," Donny said, "but the infection only got worse. Much worse." He had been applying the same treatment that Fleming had found so problematic. Antiseptics didn't work then, and they weren't working now. "The pain got so bad I had to come in."

"I'm glad you did."

"A paper cut! Can you believe it?"

"I can. Let me take another look." I asked Donny more about his exposure to benzene. He had been subjected to all sorts of hazardous chemicals; how had his doctors fingered benzene as the culprit? Donny thought it had something to do with the appearance of his bone marrow under the microscope but wasn't entirely sure. Once the leukemia diagnosis was made, his doctor recommended chemotherapy and, eventually, a stem cell transplant. "She said to get a Mediterranean transplant. Imagine that. Didn't even know that was a thing."

"Is that something you can request?"

"I'm Greek, so I suggested my sister. She got the cheek swab, and guess what? A perfect match!"

"That's great," I said, still staring at his finger. "And the transplant went okay?"

"My blood likes my sister," he said, "but my skin doesn't." Donny was once again scratching his forearm. "Got something called graft-and-host. It's like my sister's way of gettin' back at me." The stem cell transplant had been a success—his sister's bone marrow had successfully engrafted—but the new marrow was attacking its host. Under his skin, the foreign cells were fighting the old ones, releasing a burst of molecules that caused Donny to itch uncontrollably. "It's brutal," he added. Graft-versus-host disease can be lethal—several of my patients have died from it—but I didn't mention that. I reminded him that he was in good hands.

"Amazing Things Are Happening Here!" Donny quipped, lightly mocking our hospital's motto.

Eventually, after talking about stem cells and benzene and the Zadroga Act and Jon Stewart, then the longtime host of TV's *The Daily Show* and a vocal proponent of the bill, I handed Donny the consent form. "You would sign here," I said, choosing my words in a way that would easily allow him to say no.

He smiled and put down the form. "I'll take the drug, I'll sign the papers. Whatever you want."

I explained that at this phase I was simply gathering information and not yet administering the drug. "We're just observing. I want to see what happens to you."

"Me, too."

"I mean—"

He waved my words away and smiled. "I know what you mean." Donny signed the form and handed it back to me. "I'm getting this chemo drug," he added. "Starts with a *b*." He was referring to Bortezomib. "Costs three thousand a month, and they're covering it. Government's covering it. Can you believe that?"

"As they should."

"We're dropping like flies," he said, looking at his engorged finger. "Fire. Police. State troopers. My only regret is putting my wife through this." As he spoke, I thought of my own family: my wife and two small children. Would I sacrifice as much as Donny had? Would I have been scrambling around my retirement home for my white coat and stethoscope after a terrorist attack?

Donny wasn't the only person I knew who had leaned into danger that day. Tom Walsh had taken the 3:15 a.m. train from DC on September 11 to attend a morning drug discovery conference at the Stanhope

Hotel on Eighty-First Street and Fifth Avenue. A bit before nine o'clock, someone wheeled a television set into the conference and interrupted the lecturer. "You all need to see this," he said.

An hour and a half later, when Tom watched the second tower collapse, he stood up and said, "People need help. Who's coming with me?" He rushed outside with three colleagues and convinced a bus driver to take them downtown. A half hour later, he was commanding a makeshift emergency room near New York University Medical Center, treating patients with smoke inhalation, eye abrasions, and broken bones. He and Donny spent the aftermath of 9/11 just miles apart, trying to bring order to chaos.

When he was done scratching, I shook Donny's hand. "We owe you so much," I said. "Thank you for doing this."

"You know," he said as I headed toward the door, "if I hadn't grabbed my boots and gone down there, if I had just stayed at my house . . . I never coulda lived with myself. I woulda died of shame."

Chapter 15

Remy

After I stepped out of Donny's room, I felt a buzz in my pocket: a text message from Tom Walsh. "Find me. It's urgent." My mind raced as I tried to imagine what this might be about. One of his patients? Or his family? Was *he* sick? Tom pushed himself so hard that occasionally his body simply broke down. I'd seen him hospitalized before—I thought he was going to die from pneumonia shortly after we began working together—and I said a brief prayer as I jogged across the ER and up four flights of stairs to his office, recalling how from his hospital bed he'd joked that if anything was going to kill him, he wasn't going to let it be an infectious disease. I knocked and, before receiving a response, flung open his door. He was seated at his desk, his hands over his mouth. "What's up?" I asked.

Tom looked up from his computer screen and said, "Check your email."

This circuitous route to pass along critical information—text to conversation to email—was typical. I glanced down at my phone and saw that he had just forwarded something to me. As with his text, the subject line read "Urgent."

A family in Germany had reached out to Tom to help with their ailing daughter, Remy, who had contracted a fungal infection. The girl had been diagnosed with acute leukemia, just like Donny, and she had just completed her fifth round of chemotherapy. The treatment was working: it was wiping out the cancer cells, but it was also destroying her immune system, leaving her vulnerable to infection. It was an anticipated side effect of the medications, and potentially a deadly one. She, too, could become gravely ill from a paper cut. In the email, Remy's parents explained that their daughter had developed a sharp pain in her back and was unable to urinate; an MRI (magnetic resonance imaging) scan showed a large mass, and a spinal biopsy revealed *Saprochaete clavata*. The fungus was spreading, and her doctors didn't know how to contain it. "Thank you for agreeing to help us, Dr. Walsh," the email concluded. "We will do whatever it takes to save our daughter."

Fungal infections of the spine happen to be our specialty—Tom and I have published extensively on the topic—so it was no surprise that the family had been referred to him. But this organism was bizarre; I had never heard of it and had no idea if Tom had, either. It had never been mentioned in my years of medical training, and I wasn't sure it was treatable. I lifted my eyes from the screen and met Tom's. "Strange," I said.

Tom winced, and I could tell that the girl's illness pained him physically. He absorbed his patients' agony and their parents' grief in a way most clinicians did not, even from the other side of the world. As I reviewed the email, Tom gave me a primer on the infection. I should have known he would be familiar with it. *Saprochaete clavata* had once been an uncommon cause of disease in humans, he explained, but it had recently emerged as a stealth menace, especially in patients with immune impairment. Not surprisingly, no pharmaceutical company was interested in developing a targeted treatment. Perhaps most alarmingly, it was spreading: between the fall of 2011 and 2012, thirty cases occurred in France.

"That's not normal," I said, trying to think of a unifying risk factor. "I wonder . . ."

"Almost all of the patients had leukemia," he said.

"An outbreak?"

We both knew the implications of invoking that word. Fungal outbreaks were rare, and when they happened, our team jumped into action, scrambling for resources to perform our experiments. In short order, we could try new drugs in test tubes, rabbits, and humans. But the work was expensive, and we would need someone to pay for it.

The first project we had collaborated on involved the outbreak of fungal infection in brains and spines all over the United States. The culprit was a mold, *Exserohilum rostratum*, with no established treatment, and my assignment was to figure out which drugs might work. As I toiled away in the laboratory in the fall of 2012, we discovered that thousands of patients had been exposed through contaminated steroid injections that had been administered to relieve chronic back pain. Hundreds developed meningitis across twenty states, and sixty-four patients died. A negligent pharmacist at a compounding pharmacy in Framingham, Massachusetts, was eventually blamed for the tainted steroids and tried for murder. (He was acquitted.) "Sounds like an outbreak," I said again. "Something we should investigate."

Tom and I had written about the *Exserohilum* disaster in the *New England Journal of Medicine* and highlighted the need for better treatment options. There were just three major classes of antifungal drugs, we noted, and no new classes had been approved in years. There's just not much interest in finding a cure for something so rare. Remy's case was similar: she had a fungal infection for which there was no established remedy. "Let's call the family," Tom said. "I want to talk to Remy."

"I also have the doctor's contact info," I said, reading it from the email.

"Ready when you are."

We soon learned that Remy had been taking the prophylactic antibiotic ciprofloxacin to prevent bacterial infections. It was the same drug Donny had been prescribed, and while it had impeded certain forms of bacterial infection, it didn't stop all of them. It's one of the most common antibiotics in use—many people take it on vacation to prevent traveler's diarrhea—despite the fact that its chemical structure is nearly identical to that of Omniflox, the lethal antibiotic that was pulled from the market. Cipro is everywhere; it's a reliable drug that physicians all over the world rely upon. We use it because it works, and that's the problem.

Its most egregious use has been in farm animals, where it helps to

grow bigger and allegedly healthier meat and poultry. The indiscriminate use of antibiotics in animals has been one of the primary drivers of
superbugs. Bacteria living within animals get exposed to our best drugs
and learn how to avoid them. A recent outbreak that affected more than
one hundred people in eighteen states was ultimately linked to an unexpected culprit: puppies. Nearly all of the infected dogs were sold at pet
stores and had received at least one course of antibiotics, and the lethal
superbug was passed to their new owners.

As Remy's doctors spoke, I wrote "Stop the Cipro?" on my notepad
and pushed it in Tom's direction. He held up a finger, indicating that he
wanted to hear more about her case. It was unclear how Remy acquired
her infection, but we soon learned that she had been taking Cipro for
four months. As expected, it had protected her from bacterial infections,
but she was still susceptible to viral and fungal infections. Over the
next forty-five minutes, the German doctors and Tom and I discussed
whether to continue with Cipro now that a fungus had been implicated.
During a pause in the conversation, I pulled up flights from New York
City to Munich. Tom opened his calendar. As we batted around ideas for
treatment, Remy's anxious doctors kept returning to a single question:
Why did this happen?

CHAPTER 16

A Quiet Revolution

THE ANSWER INVOLVES a fundamental change in how we treat leukemia and other forms of cancer. For the past few generations, doctors have used a combination of surgery, radiation, and chemotherapy, and in the push to save lives these interventions have become ever more aggressive. I routinely care for patients who are receiving treatments far beyond what was standard the previous year. Like bacteria, cancer cells can develop drug resistance, mutating in ways that neutralize and inactivate our best drugs, and the toxicity associated with chemotherapy leaves some patients so ill they're unable to continue with treatment. An overnight doctor in my hospital is often called to the bedside of a vomiting cancer patient in urgent need of rehydration. We're pushing patients further than we ever thought possible, offering them a slim chance at a cure.

Remy's bone marrow was responding to harsh chemotherapy, but it was putting her sixteen-year-old body in uncharted territory. The cancer treatment had wiped out her white blood cells, which serve as the last line of defense against infection, and microbes in the environment seized on her body as an energy source. During the conference call, we learned that there were signs that the toxic therapy might soon stop working, and her doctors were scavenging for other options to save her.

One option is immunotherapy, which uses the patient's own immune system to destroy cancer cells. White blood cells can recognize cancerous, misshapen proteins and carbohydrates as *foreign*, and scientists have figured out how to harness these powers of detection to attack the malignant cells. This approach has revolutionized our expectations for dying patients, and the scientists who pioneered its use were awarded the Nobel Prize in 2018. People are living longer because of immunotherapy—it has added several years to the life of former president Jimmy Carter, who announced in 2015 that he'd been diagnosed with advanced malignant melanoma—but the new treatments can also cause the immune system to go haywire. The human body's response to infection is tightly orchestrated, and any disruption—even one designed to fight cancer—has the potential to weaken its reserve. That leaves an opening for superbugs.

Remy's aggressive treatments had left her susceptible to infection, and any further disruption to her immune system could be fatal. Immunotherapy might save her life, or it might kill her. Clinicians and families around the world were making the agonizing decision to proceed with a new treatment, leading to more and more frantic phone calls to Tom Walsh, who happens to be an expert in both cancer and infectious diseases. Bacteria and fungi and parasites that weren't previously known to cause disease are suddenly in play, and doctors aren't sure what to do. So they turn to Tom.

As he gave his instructions to Remy's doctors, I took notes. When he hung up the phone, I shook my head. "I've got a new reading assignment," I said, pointing at the words *Saprochaete clavata*. I would spend the remainder of the day and the following week scouring the library to find every paper ever written about Remy's infection. As it turned out, there wasn't all that much to read. Remy's physicians had already dissected those unsettling papers and understood that they needed an expert to guide them. I had a distinct takeaway from the literature: Remy would soon die from her infection.

The clinical atmosphere was chaotic, with dangers both seen and unseen swirling all around us. We were playing whack-a-mole, devoting resources to one unfamiliar bacteria while others were left unattended, stealthily gaining strength and agility. At times, it felt like we'd been looking in the wrong direction, ignoring the threat from within—focusing on ways to grapple with howitzers and tanks, when an equal danger was posed by the spores of tetanus beneath the soil.

Tom and I spent many long nights trying to make sense of it all, writing grant proposals and lectures that would provide cover from the gathering storm. When he was excited by an idea or agitated by a potential nuisance, Tom would put on Beethoven and pace around his office, tossing off ideas that I scribbled down furiously, trying to keep up. Those soliloquies often took me days or even weeks to process, and they formed the foundation of our small contribution to the fight against superbugs. Were we making progress? On most days, I couldn't tell.

Tom Walsh

Mornings were spent playing catch-up, slogging through medical journals and press releases from the FDA and drug companies to understand how the landscape was shifting. Some days the news felt monumental, like when the NIH announced that it was going to fund nontraditional approaches to fighting superbugs, but most of the time the changes were subtle, and often reached us as a tiny ripple from another field of medicine such as rheumatology, oncology, or hematology, the science of blood disorders. A shift in the approach to treating leukemia, for example, might have far-reaching effects for those trying to prevent and treat infections.

While cancer immunotherapy is risky, it often works. And, crucially, it's reversible. What works for Remy, however, might not work for Donny, and I've treated a handful of patients who have developed complications from a single pill. This flexibility has its advantages—I can pump the brakes when something clearly isn't working—but it has pitfalls, too, and some oncologists have been pushing for a more permanent cancer therapy, one that would alter our genetic code itself: CRISPR. It was something that came up time and again as I searched for a better treatment for the girl from Germany.

CRISPR, an acronym for *clustered regularly interspaced short palindromic repeats*, is the most important scientific discovery of the century, because it allows for the permanent modification of DNA. Problematic genetic material can be removed forever: for example, cancer-causing genes can simply disappear. It's an amazing advance, but there are technical and ethical hurdles. Permanently altering the genetic code can change what it means to be human; it wouldn't be *that* hard to insert rabbit DNA into a person, and we could do it without someone's consent. Investigators are also exploring CRISPR to fight superbugs: we could simply cut out resistance genes and weaken pathogens to the point that they no longer pose a threat. Remy's doctors wouldn't need to call Tom Walsh; they'd simply devise a CRISPR message for *Saprochaete clavata* to self-destruct. But that type of solution was years, if not decades, away.

In the more immediate future, CRISPR may be used in a slightly different way. After I learned about Remy's condition, I came across the work of Timothy Lu, a synthetic biologist at the Massachusetts Institute of Technology (MIT), who was using CRISPR as a diagnostic tool to detect superbugs and as a therapeutic tool to precisely kill them. "The idea," he told me, "is to use it as a molecular scalpel. We can use CRISPR to find drug-resistant *E. coli* quickly, homing in on a single mutation in a key enzyme such as DNA gyrase, to help doctors know what antibiotics might work." This is the future of precision medicine, he said, using molecular tests to differentiate the most dangerous pathogens from those that are harmless, so doctors can select the best treatment. It might be a way to neutralize Remy's infection.

AS WE DISCUSSED treatment options for Remy, I learned more about the intricacies of her case. After developing a fever, she had been given an antibiotic called meropenem, which failed to halt her rising white blood cell count, dropping blood pressure, and other ominous signs of infection. "I'm not surprised," Tom said, shaking his head. "Mero's not the right choice here." Meropenem is one of the most commonly used antibiotics for severe infections because, in contrast with most other antibiotics, it is highly resistant to enzymatic degradation. Bacteria have been outsmarting our best drugs, but not this one. In the middle of the night, when I'm searching for the right treatment for a sick patient and I'm unsure of the diagnosis, I often turn to meropenem. It's safe, it's effective, and I can usually convince the antibiotic steward—the physician or pharmacist who dispenses antibiotics and prevents their unnecessary use—to give it to me because it's not *that* expensive. Bacteria quickly learned to produce enzymes to chew up other antibiotics, but for years, meropenem was relatively impervious and a reliable standby—until recently.

Chemical structure of meropenem

The problem began in November 2007, when a fifty-nine-year-old Swedish man traveled to India. Soon after arriving, he was hospitalized with an abscess near his pelvis and was transferred to a hospital in New Delhi for surgery. Diabetes was thought to contribute to the infection—elevated blood sugar impairs the immune system—and several weeks later, the patient developed a urinary tract infection. All of this was rather mundane; doctors had no reason to suspect anything unusual was afoot, and he eventually returned home to Sweden. But the bacterium isolated from his urine was something that doctors had never seen before: it possessed an enzyme that could destroy meropenem. It was a disastrous development—a signal that bacteria were once again winning the tug-of-war against patients—and the news sent researchers like Tom scrambling back to the drawing board. If bacteria developed full resistance to meropenem, tens of thousands would die each year.

The deadly enzyme was dubbed New Delhi metallo-beta-lactamase-1, or NDM-1, and no one really knew how it appeared or where it might spread. NDM-1 is carried on a small bit of DNA called a plasmid that can pass easily from one bacterium to another. The enzyme was going to mobilize, but we didn't know where. As the authors of the first NDM-1 paper concluded, "In a country where there is little control on antibiotic prescriptions, the rapid dissemination of such a plasmid is alarming." NDM-1 reached the United States three years later.

When Tom and I reconvened to discuss Remy's case, I learned that the abscesses in her spine were expanding, and she was unable to walk. She'd soon become paralyzed. Her doctors were nervous, and so was I. "I have an idea," Tom said. He laid out a strategy that would rely on three antifungal agents, given at high doses that few doctors would ever try. It was a Hail Mary, a last-ditch effort to save a young girl a world away.

"The drugs might antagonize each other," I said, noting that some antifungal drugs can weaken the effects of others.

He shook his head. "They won't." Tom Walsh did not bill for his time or his expertise. It was all part of the mission. "We'll get her through this," he said, with more confidence than I could muster. Tom examined the cartoon characters on his necktie—something a pediatrician would wear—and looked at me. "I have faith." He closed his eyes and grimaced. Once again, I could see him absorbing the pain of his patients. I wondered how Remy had acquired leukemia. Was it a random mutation or exposure to a toxic chemical such as benzene? I'd probably never know. Tom grabbed his white coat and patted me on the back. "Let's get to work," he said.

I had spent the early part of my career trying to see the world as Tom did. In his presence, time could expand and contract. There were agonizing stretches where I watched and waited as he drew up equations that only he understood, followed by fleeting moments where he made an impossible idea accessible, or provided a solution to a problem I'd given up on. Time is of the essence, he often said, but in his world, it was just an abstraction. Things that seemed irreconcilably diffuse could be brought into focus, but it took effort: endless hours at the laboratory bench and at the chalkboard and in the lonely corridors of the hospital.

At times, it was intimidating to collaborate with someone so wholly focused and determined. To leave the laboratory early or to skip a scientific meeting might imperil the mission and set back our work. Tom carried ferocious responsibilities—Remy was just one of many he managed from a distance—and when I was around him, I felt the weight. He called Remy's doctors and told them how to proceed. "Let's talk again in a few hours," he said. "Keep me updated." As Tom spoke, I realized that the plan he had just come up with had changed slightly. He was charting yet another course for the girl. I glanced at the reminder on his desk: "We Defend the Defenseless." The words gave me hope. They elevated our work into something noble.

I had been inspired to pursue this career by the Harvard physician Paul Farmer, the infectious diseases specialist whom the Pulitzer Prize–winning author Tracy Kidder said would "cure the world." That description also applied to the man next to me.

CHAPTER 17

Decision Points

As the dalba delivery date approached, word spread around the hospital that a new drug was coming. Despite having one of the most robust pharmacies in the country, NewYork-Presbyterian Hospital didn't carry anything like it. In meetings, in elevators, or just walking to the cafeteria, curious clinicians would stop me and ask for an update. When would dalba arrive? How would it change things? And what took so long?

Introducing a drug into a major hospital for a clinical trial is no small feat. Any study drug—from a cheap blood pressure pill to an expensive antibiotic—is shipped to the hospital's investigational pharmacy, which is responsible for subject enrollment, order entry and verification, and preparation and dispensing. Once it arrives, staff receive training in storage and administration, and a physician hoping to use the medication must be added to a list of approved providers. After that, she must submit a request to the pharmacy along with the signed informed consent and associated registration documents for the patient who will receive it. Dalba would be closely guarded; ordering it would be like requesting a security clearance.

Once a patient has agreed to take the drug, the pharmacy dispenses it to the patient's nurse. After it is administered, the patient is monitored closely for signs of an adverse reaction, which could range from itchiness or hives to Stevens-Johnson syndrome. It's a tedious process, but a necessary one. The system works.

Once a new drug such as dalba is in stock, another gatekeeper enters the picture. If a doctor wants to prescribe an antibiotic, he must get permission from an antibiotic steward, who reviews the case and rejects requests that are deemed inappropriate. This adds another layer of complexity and potential confusion to the doctor-patient relationship. Your physician might tell you she's prescribing ciprofloxacin for a urinary tract infection, only to find that the steward has recommended amoxicillin instead.

Antimicrobial stewardship programs have been around for a few decades but have taken on a much larger role in patient care in the past few years. Stewards tend to be infectious disease specialists or pharmacists, and their role is to promote the appropriate use of antibiotics, which decreases the spread of superbugs. The next time you're prescribed an antibiotic in the hospital, you should ask if a steward approved it.

During my training, I moonlighted as an antibiotic steward, advising doctors on nights and weekends while restricting access to coveted antibiotics. It's a thankless job—most of the time, you're dissuading a doctor from doing something she really wants to do—but it's a necessary check that saves the hospital tens of thousands of dollars. More importantly, it spares patients from exposure to unnecessary drugs.

Decisions about antibiotics are not binary: it's not just a yes-or-no question of whether to treat an infection. When antibiotics are given, there's an aggressive push to prescribe shorter courses. For example, in medical school, I used to treat bacterial pneumonia for eight days; in residency, it was seven. Now I treat most pneumonia for five days, and a recent study suggests that we can go as short as three days. Stewards have implemented these changes, preventing millions of unnecessary prescriptions from being written.

There is, however, an unintended consequence of these efforts: stewards curb investment in antibiotics. Drug developers know that their expensive new drugs will be highly restricted at most hospitals and that doctors will have to request (or beg) to use them. It's far more lucrative for a company to create a drug that everyone can use, day after day, for years, without asking for approval.

Kent Sepkowitz, an infectious disease physician at Memorial Sloan Kettering Cancer Center, in New York, astutely described the changing nature of our field: "Once, we were the unwavering champions of antibiotics," he wrote in the *New England Journal of Medicine*, "imploring our colleagues to give them early and often. But increasingly, we now find ourselves on the other side of business, as the official chastisers about overuse and abuse. We have come to adopt an almost Victorian attitude of abstemiousness." It's hard to look someone in the eye—a nervous physician or a vulnerable patient—and say no, but increasingly that's what our job has become.

I knew that dalba was just one piece in the superbug puzzle. We needed more treatment options, and as my study enrollment continued, I courted other companies with new molecules that needed testing in humans. This was the essence of *translational* research, a buzzword in academic circles reflecting the need to translate discoveries in the laboratory into meaningful advances for patients. I was a translator, perpetually scavenging for text.

Selecting the right one is tricky: what looks good on paper can be dangerous in practice. But I was delighted by the hunt; learning about a drug and then digging deeper and deeper into the experimental work was the joy of being a doctor. I had the power to bring new medicines into one of the best hospitals in the world, to find out what worked and what didn't. Exposing strangers to unproven products was a delicate task, and it forced me to engage with medicine and science in a way that I never had as a student or as a trainee. Defending the defenseless was perilous work, and when I entered the hospital each day, I was reminded just how easy it would be to screw it up. Finding the next lifesaving drug kept me up at night, and it put a charge in my step when I put on my white coat.

One afternoon Tom identified a candidate: an antifungal drug so new that it hadn't yet been given a name. The drug was discovered at

Merck at the turn of the twenty-first century, but the company abandoned it in 2013 following a disappointing portfolio review (a detailed
financial analysis of an investment), and licensed the compound to a
New Jersey–based company called Scynexis. Early studies indicated that
it was effective in a test tube—it kills just about every fungus in sight—
but Merck wasn't certain it would work as well in humans or that the
market could sustain its hefty price tag. The pharmaceutical giant struc-
tured the deal such that it was still entitled to royalties if the FDA ever
approved the drug, but it's unclear if and when that would ever happen.

After granting the coveted designation "orphan drug," which provides seven years of market exclusivity to a drug that addresses a rare
disease or treats a common condition but is not expected to generate a
profit for the manufacturer, the FDA told Scynexis not to start trials of
an injectable form of its drug after several healthy volunteers developed
blood clots in a phase 1 study. (Phase 1 studies typically involve fewer
than one hundred people; phase 2 studies usually include several hundred patients; phase 3 studies may have several thousand.) It was a co-
lossal blow, but the company was undeterred; it suspected that the real
promise of the drug was in the oral formulation, which could be given
for an extended period of time while shortening hospital stays.

Tom produced a study protocol and asked me to review it. After
hammering out the details—given the extensive follow-up requirements, I insisted that we pay participants handsomely—we submitted it
to our hospital's Clinical Study Evaluation Committee. CSEC was another gatekeeper of sorts, staffed by clinical researchers and statisticians
who could determine both the scientific merits and feasibility of any
study. It was the first hurdle in the process, and it was required before
submission to the institutional review board. We were optimistic that
our study would be approved—the *in vitro* work with this antifungal
drug was superb—but the blood clots still concerned me.

"I'm not sure what CSEC will think," I said to Tom as we walked into
the committee meeting. Unlike IRB meetings, we could attend CSEC
conferences to defend our proposal. "I'm ready to fight." By that time, I

had enrolled dozens of patients in the dalba study and was eager to take on additional projects. This felt like the right one.

An oncologist at the head of the conference table called the meeting to order. He was wearing a somber grey blazer and had a stack of papers next to his coffee. There were a dozen or so men and women seated around the table, munching on muffins and sipping coffee as Tom introduced the study. "This is a phase two trial," he said, "to test a new molecule." He explained how dire the situation had become: patients with fungal infections were no longer responding to treatment, and their infections were coming back time and again. With every recurrence, they became more difficult to treat. "Mortality approaches eighty percent," I added, thinking of Remy, "and this drug offers them hope." Our trial would involve men and women for whom all of the available treatments had failed. Only sixty patients would be recruited, but they would be evaluated extensively, with at least a dozen scheduled follow-up visits and blood tests for every enrollee. And they would be closely monitored for blood clots.

When Tom was finished speaking, the oncologist cleared his throat. "First of all," he said, "thank you for that overview. But there are some issues here. First, there is no predefined endpoint." I looked around the room to see if others agreed. A few heads nodded.

"I disagree," Tom said.

The oncologist looked at the papers in front of him. "In my field, we predetermine efficacy. If x percent of patients respond to a treatment, the drug works. If not, it doesn't." He pointed to his papers. "This trial doesn't do that."

"That's correct," Tom said.

"Why?" the oncologist asked.

The two men stared at each other for a moment as we all looked on. "Let me make something clear," Tom said, craning his head around the room. "This is exploratory. This study is different because *these patients are different*. They are desperate."

"But there's no power calculation," another man said. "It's a single-arm study. You're not comparing this new molecule to anything. Why?"

"We have no options," I said. Heads turned in my direction. "There's nothing left to compare it to."

Tom nodded. "It's true."

"Other thoughts?" the oncologist asked. He waved his hand over the table, indicating that the floor was now open for comments. The next few minutes would determine the immediate fate of the study. We might never make it out of committee. There was something perverse about specialists in other fields questioning our reasoning, but this was done to ensure objectivity. I felt a quiet anger radiating from within as I watched the men and women review the protocol. "We're out of drugs," I said finally. "And these patients are out of options."

The room was silent for a moment. I couldn't tell which way things were going to go. Then a woman in the corner spoke up. "I get it," she said. "The endpoint is different here. The entire situation is different. We should approve the study." The oncologist looked at his papers and then at Tom. "I agree."

Tom stood up. "Thank you," he said, and walked out of the room. I caught up to him at the elevator.

"Success," I said.

We stepped out of the elevator and walked to Sixty-Ninth Street, where Tom hailed a taxi to the airport. "The mission continues."

CHAPTER 18

Piper

TOM DEPARTED FOR CHICAGO, and I returned to my office. I was nearly done with the first phase of the dalba study and needed to enroll just one more patient. Sifting through the latest round of paperwork, I reflected on the patients I had already recruited. People like Ruth and George and Erwin and Donny. People with wildly different backgrounds, lending their help for a study they knew relatively little about. A trial that might not directly benefit them; one that might not finish until after some of them had died.

A half hour later, I found a new patient for my study: a thirty-four-year old woman named Piper Larson, who had just been admitted with a painful red lump near her left collarbone. The ER physician suspected a skin infection and had ordered intravenous vancomycin. I grabbed my stethoscope and headed to her room. Piper was scheduled for an ultrasound, and I wanted to catch her before she was whisked away.

Piper had grown up in North Carolina, near where I was born in Charlotte, and she was now a photographer in the West Village but had been out of work for several months. "Once I was diagnosed," she said, "I stopped working. I had to. Everything stopped so I could focus on my health." I hadn't read her chart in detail, so I didn't know what she was referring to. All I knew was that she possibly had a skin infection. Piper handed me a thick packet of papers—health records from her primary care doctor—and filled in the gaps in her medical history. As she spoke, I read. *What is the diagnosis?* She ran her hands through her ombré-dyed hair as I turned the pages. "I'm still in shock," she said as a tow-headed boy entered the room.

He looked to be around six years old, and he was wearing Superman pajamas—the kind my son would wear—and he had probably been with his mother throughout the interminable night in the emergency room, waiting for a bed to open up. The boy had wire-rimmed eyeglasses and was missing one of his two front teeth.

"Hi, Mom."

I glanced at Piper's chest, looking for the lump, and returned to the documents, scanning until I found the diagnosis: gastric carcinoma. Stomach cancer. The boy leaned against his mother. "Can I go to the vending machine?" he asked. "Please?"

Piper had first come to the hospital several years earlier, after she developed a pain in her stomach that wouldn't go away. Like Tom Walsh's mother, she had tried to fight through it, but eventually the pain had become unbearable, and she had an endoscopy. A biopsy ultimately revealed the news. Piper's prognosis was grim, but there was hope as long as the tumor hadn't spread to other parts of the body.

"Mom?"

"In a few minutes," Piper said. "When Dad gets here."

That was the kind of thing my wife would say to my son: "Just a few minutes. We've got time." But maybe she didn't. A diagnosis of gastric carcinoma could be a death sentence. I put down my papers and just looked at the two of them: mother and child. What would he remember of all this? Would I be a part of this terrible childhood memory?

"Are you okay?" Piper asked me.

My eyes welled up. "Yes," I said, catching myself. "I'm sorry. Just thinking of something."

"Am . . . I okay?" She smiled, hoping to bring levity to an exchange that was becoming uncomfortable for both of us. "Everything all right?" I wasn't sure. It must have been awful to see a physician on the verge of tears, someone she didn't even know, but I was now wondering if the swollen lump might be a sign of metastasis, an indication that the cancer had moved from her stomach to a lymph node or bone. I wondered whether this boy might grow up without a mother.

"I just need a minute," I said. I stepped out of the room and pulled up her file. The cancer was supposedly in remission, but this lump near her neck could be a harbinger of disaster. It might not be a simple skin infection; it could be a cancerous lymph node—Virchow's node. The lymphatic system was designed to trap infections in a complex network of tissues throughout the body, but occasionally these lymph nodes captured a tumor. Or maybe it was nothing, a bug bite or a contusion after a fall. Regardless, she wasn't appropriate for my trial—not until the node situation was sorted out. It didn't look like cellulitis, and the uncertainty excluded her.

I returned to the room, sidestepping Piper's son, who was now lying on the floor, holding a lollipop. I looked at her lineless face and the swollen lump, and then I turned to her son. I tried to keep my composure while imagining the boy growing up without a mother, just like Tom Walsh—the bookish young man with a nose for science, a firm handshake, and an affinity for the military. *The Little Engine That Could.* "I reviewed your chart," I said, trying to keep it together. "You're not quite right for the study." Piper nodded and handed her son more coins for the vending machine as I left the room. "At least not right now. Thank you for your time," I added. "You may see me again."

I wandered down the fluorescent corridor, staring at my shoes, once again engulfed by the clatter of modern medicine: physicians and nurses hurriedly typing into cell phones and computers as patients and their families passed by. It was an odd place to work: at times amazing, but it could also be devastating. The wonderful parts of being a doctor—the cures, the relationships, the discoveries—were always counterbalanced by the tough moments. The encounters that leveled me. I still hadn't figured out how to prepare for them. Perhaps I never would. When I looked up from my shoes to turn a door handle at the end of the long hallway, I felt a tap on the shoulder. It was Tom. "What's wrong?" he asked. "What's going on?"

"I thought you had a flight," I said. "I thought—"

"Plans changed. What's going on?" The man who absorbed others' pain was about to take on mine. The lineaments of heartache and exhaustion were etched across his face; I knew that one day they'd be scrawled across mine, too.

"Tough day," I said, my voice cracking as the words dribbled out. "Just tough."

He put an arm around me and squeezed. "I know," he said softly. "What can I do?" Under the gentle words and the heavy eyelids, his buoyancy was unmistakable, his enthusiasm undimmed.

"I don't know," I said. "Get my mind on something else. A moment of whimsy might be nice." I laughed as Piper had, trying to lighten the mood.

Tom thought for a moment. "Are you familiar with the miniature Chincoteague ponies of coastal Maryland?"

I wiped my eyes and laughed again, this time for real. "As a matter of fact, I'm not." Over the years, I had been subjected to hundreds of his mini-lectures on topics ranging from the Peloponnesian War to perestroika. But this was new. "Tiny ponies?"

"Take a walk with me," he said.

"Thank you."

"And before I forget," he said, "I have an update."

"Yeah?"

"Remy's doing better." His face lit up. "Much better. Infection is almost entirely gone."

CHAPTER 19

Garden State

WHILE THE ANTIFUNGAL STUDY was wending its way through committee, Tom and I were busy devising other ways to hasten the drug's availability to the expanding cohort of patients in New York who were infected with a new fungus called *Candida auris*. The organism had been discovered in the ear of a seventy-year-old Japanese woman in 2008 and promptly spread all over the globe. The pattern of distribution was unpredictable—an intensive care unit in the United Kingdom found the fungus on its reusable armpit thermometers—and it was now popping up in Manhattan. The bug was often resistant to antifungal treatments, and mortality was skyrocketing. I was in Tom's office when a reporter from the *New York Times* called him and asked what could be done. "You should talk to Dr. McCarthy, too," he said before handing me the phone.

We had seen reports that the new antifungal drug could kill *Candida auris* in a test tube, and we were eager to make it available to patients. This was a dicey move. We were bypassing the standard practice of exhaustive testing, but there simply wasn't time. I had seen what happened to my patients with *Candida auris* infection—some required sequestration in a specialized room to prevent the spread of the fungus, followed by repeated surgical procedures—and was anxious to find a better option. Tom arranged a meeting with the makers of the drug, Scynexis, and traveled to its corporate headquarters in New Jersey to make our case.

Scynexis moved from North Carolina to Exchange Place, a concrete office park in Jersey City, in 2015, shortly after spinning off from Merck—and just as *Candida auris* was reaching New York. Exchange Place, which overlooks the Hudson River and the southern tip of Manhattan, is a bland square, just two hundred feet long; companies have flocked there for the cheap office space and gorgeous views, dubbing it Wall Street West, although it has none of the trappings of Manhattan's financial district.

Tom and I passed an Au Bon Pain and a row of poinsettias as we walked the short strip of concrete along the west bank of the Hudson. "Do you know where they got their name?" he asked, pointing at the flowers. "The first US minister to Mexico, Joel Poinsett, was a physician and botanist. His—" Before he could finish the thought, we were intercepted by a receptionist and ushered to a conference room on the thirty-sixth floor, where Tom was greeted like a minor dignitary. Everyone at Scynexis was familiar with his work.

"We're here to talk about a powerful new drug," a South American physician named Sylvia said as we took our seats. "We're excited that you're excited." Sylvia and Tom had worked together on other development projects for years, and she turned the floor over to him.

"We need a protocol," Tom said, standing up to drive home his point. "Cases of *Candida auris* are emerging in places we never expected, and patients are dying." He looked at me and extended his left hand, passing a metaphorical baton.

"I had a case last week," I said. "And one the week before." One of my patients had undergone more than a dozen endoscopic procedures in a frenzied attempt to excise the infection, but *Candida auris* kept spreading. After the fourteenth procedure, my patient pleaded to let her young daughter visit her, but I had denied the request: it was simply too risky. The woman had finally been cured when Tom took on the case.

"We've reached a breaking point," I said to the team at Scynexis. "We need something new to offer. And we believe you have it."

Sylvia pulled up a PowerPoint slide. "We're here to help," she said, pointing at the screen. "We've developed a protocol to fight this." She spoke with dispassionate authority, briefing us on the details of the study. It was rigorously designed, but I knew it would take months to receive CSEC and IRB approval.

"We need something faster," I said. "What about compassionate use?" It was a mechanism to get unapproved drugs to dying patients when there were no other treatments. I scanned the room to see how the idea was received. The faces were mostly blank. "We can get it to people quickly, and we can circumvent the regulatory stuff."

A man at the head of the table shook his head. "We need a protocol," he said. "We can't just give the drug away."

"I agree," Tom said. "We need to know who benefits and who doesn't. It needs to be studied."

"That takes time," I said. "Time we don't have."

"I agree with Tom," Sylvia said. "We need to do this the right way. And that'll take some time. We already have this protocol up and running in India." The subcontinent had been hit hard by *Candida auris*, and its regulatory framework was radically different from ours. Rapid distribution was a different sort of challenge in India. "With your help," Sylvia went on, looking at Tom and then at me, "we'll have an emergency-use process here, too."

"Eighty percent of the cases in the United States," I said, "are in the greater metropolitan area. And not just big hospitals." I looked around the room as heads nodded gently. "We need to capture everyone," Tom said. "The small hospitals, the nursing homes. Children's hospitals. Everyone." *We defend the defenseless.*

"Which brings me to my next slide," Sylvia interjected. It included a picture of a telephone. "We need a hotline," she said, "and we need to define responsibilities."

"We'll handle it," Tom said. The knee-jerk response didn't surprise me. From hurricane relief to holiday coverage at the hospital, he never passed up a chance to volunteer. Before moving from the NIH to Cornell, the former altar boy and Eagle Scout had risen to the rank of captain in the US Public Health Service, jumping at the chance to join the Commissioned Corps Readiness Force to provide disaster relief everywhere from post-Katrina Louisiana to Lower Manhattan after 9/11. "We'll do it," Tom said, looking at me.

"This is twenty-four/seven, three hundred and sixty-five days a year?" I asked. If Tom was away at a conference, I knew who would be responsible for the hotline: me.

"Yes," Sylvia said. "When we get a call, we need someone ready to spring into action. Someone who can review the case, determine if the patient is infected with *Candida auris*, and get the drug to the patient. No delays."

I imagined responding to a frantic call in the middle of the night. Most physicians had never heard of *Candida auris*, and those who were familiar with it had seen only one or, at most, two cases. There was still a lot of uncertainty regarding the containment and treatment of the infection, and life-altering decisions needed to be made immediately. Complicating matters, hospitals were using different diagnostic techniques, with some relying on gene sequencing—looking for characteristic stretches of DNA to find the organism—while others used biochemical tests; the fungus was often misidentified. Our hospital employed a technology called matrix-assisted laser desorption/ionization time-of-flight mass spectrometry—a platform that most doctors weren't familiar with. Answering that call would not be a simple task. It would entail a frantic conversation and incomplete information, with someone's life hanging in the balance.

"What about vacations?" a lightly bearded man asked. "Then what?"

"I'll man the phone," Tom said. He took a sip of his coffee and looked across the table at me. "Matt and I will handle it." I had spent years observing the way Tom managed emergencies—the crashing patients, the frantic calls from doctors in far-flung places—and wondered if I could muster the same response: calm and deliberate, confident and reassuring. Tom has a quality that is as valuable as it is scarce, one that gets tossed around a lot but is truly remarkable when you find it: he is brave. He volunteers for things that I would never dream of doing, and he gets me to do things I might not attempt otherwise. This was one of them.

"We'll handle it," I said.

I wanted to be like Tom, but I wasn't convinced that I'd ever get there. I tried to remember the times I had performed well under pressure; a time when I had met a challenge head-on and succeeded. Pitching in a tied game with the bases loaded came to mind. "We got this," I added.

Our meeting with Scynexis laid bare what was so confounding about drug development. The intravenous formulation of its drug was so dangerous that the FDA halted human testing, while the oral version was so effective that we were creating a hotline to distribute it more rapidly. We spent the remainder of the afternoon ironing out logistics.

The negotiation around that conference table in Jersey City was familiar terrain for Tom. He has cared for patients from twenty-eight countries and more than a hundred hospitals, and had arranged emergency release of drugs for countless patients, many of them children. Working with Merck, he helped get the antifungal drug caspofungin to critically ill babies who were suffering from a deadly fungal infection in Costa Rica. (Eight of nine survived.) Addressing *Candida auris* was simply the next project, all part of *the mission*. "My phone is always on," Tom said.

"Oh, yes," Sylvia said, chuckling to herself as she turned off the presentation and took a sip of water. "I know all about your phone."

CHAPTER 20

Trojan Horses

"GOOD MEETING," I said to Tom as we stepped out of Scynexis head-quarters and into the fading sunlight. "Hopefully not our last trip to Jersey City." My head was swimming with ideas, dreaming up trials and collaborations, wondering if we could get this new antifungal drug to Remy in Germany, and growing more comfortable with the idea of manning the twenty-four-hour hotline. "Progress," I said softly. "Finally."

On the trek back to Manhattan, I received word that an antibiotic called lefamulin had beaten expectations in a pneumonia trial, suggesting it had far broader use than anyone expected. The drug had been fast-tracked by the FDA in 2014, but I hadn't anticipated that it would work in the lungs. (Like dalba, it was initially intended to treat skin infections.) Lefamulin causes bacterial protein production to malfunction, like inserting Lucy Ricardo onto an assembly line at the chocolate factory, but companies have had trouble finding the right formulation to make it work in humans. It had been shelved for decades, but a Dublin-based company, Nabriva Therapeutics, figured out a way to harness its potential. "It finally feels like the wind is at our backs," I said to Tom as we emerged from the Holland Tunnel.

He pointed at the deadlocked traffic and smiled. "Almost."

Lefamulin was just one in a string of successes. The FDA had also approved a new drug combination, meropenem-vaborbactam, to treat complicated urinary tract infections, including superbugs that had developed resistance to meropenem. The strange thing was that vaborbactam was useless on its own—everyone knew it had no promise as an antibiotic—but it made other drugs more powerful. I liked to think of it as an anabolic steroid, turning a slugger like meropenem into a Hall of Famer. The approval of meropenem-vaborbactam, known together as Vabomere, was the most encouraging news to hit antibiotic discovery since that missionary disinterred Mississippi Mud in Borneo. My mind leapt to the studies we would soon perform.

Chemical structure of vaborbactam

The other cause for excitement was the development of an antibiotic called cefiderocol that had been shown to kill *Acinetobacter baumannii*, one of the superbugs that routinely pop up on lists of the world's most dangerous bacteria. Cefiderocol isn't the first to treat the infection, but it is one of the first to do so using molecular trickery. We know that bacteria love iron—they have special mechanisms to scavenge for it—and the makers of cefiderocol exploited that to make a nimble new drug. They attached an antibiotic to a molecule that binds iron, fooling microbes into ingesting it. We call it the Trojan horse approach, and cefiderocol proved it could work.

"All great news," I said later that week as I reviewed the cefiderocol data with a representative from the company that makes it. "Very impressive." I was back in my office, watching flags billow in the wind. "How much does cefiderocol cost?" I asked. "Whatever it is, cut it in half." I had taken the meeting because I was interested in studying the drug and, more importantly, because I thought it might treat Jackson's infection.

"Unclear where we'll set the price point," the rep said. The drug wasn't yet approved by the FDA, so a price hadn't been determined. "But we're working on it," he said, handing me a stack of diagrams. "As you know, the typical answer is: whatever the market can bear." I had heard that phrase so many times, it had been bleached of any meaning. *What the market can bear.* More like: *Whatever we can get away with.* Patients needed these drugs, but the market wasn't designed to sustain them. Who was going to pay a thousand dollars a pill? Or more? I knew that Jackson could not.

Manufacturers typically get twelve to fifteen years of market exclusivity before generics can compete—unless the patent is transferred to an Indian reservation—but if generic manufacturers don't bother, prices can actually *increase* after the patent expires. Between 2013 and 2016, one in ten antibiotics experienced a 90 percent price hike due to lack of competition. Unless other companies create more Trojan horses, the price of cefiderocol could surge. "We won't use it if it's too expensive," I said.

The people who study this stuff usually call on lawmakers to step in and stop the price-hiking madness, but this rarely happens. Few are interested in interfering with the marketplace or allowing importation of off-patent drugs from other countries, which means that even the most groundbreaking discoveries have to be met with caution. Nonetheless, I left the meeting feeling encouraged. The pharma rep was clearly passionate about research and offered up several methods to fund a new study. A welter of ideas emerged after that meeting as I examined the data from one successful trial after another, including my own. I had enrolled the final patient in the preperiod and was ready to give dalba to my first patient.

Over the next few days, Tom's office served as a mini–think tank, and we spent long hours at the drawing board trying to predict how dalba would affect patients. This had always been the allure of working with him: anything seemed possible. He was a modern-day Horatio Alger, overflowing with enough optimism to turn a cynic like me into a be-liever. The encouraging results of lefamulin, vaborbactam, and cefidero-col were reason enough to celebrate, but they paled in comparison to an extraordinary discovery that had quietly occurred just two blocks away from our offices, across Sixty-Eighth Street at The Rockefeller Univer-sity. It was something that no one was talking about yet, no one had even heard of, but the finding would change my entire approach to the fight against superbugs.

PART 4

Beneath the Surface

CHAPTER 21

The Rockefellers

WILLIAM ROCKEFELLER SR. was a huckster, a peddler of bogus medications who occasionally pretended to be disabled to manipulate clients. Despite having no medical training, Doc Rockefeller, as he called himself, claimed to be a cancer specialist, hawking potions and elixirs to sick patients across the country. Raised in the shadow of fraud, his son John Davison Rockefeller went in the opposite direction, becoming an evangelical Baptist and captain of industry—a ruthless oil magnate who would become the richest and, at times, most loathed man in the United States. History would tell us that the Rockefellers didn't defend the defenseless, they exploited them, extracting every dime they could from people of modest means—men and women who simply wanted to heat their homes, feed their families, and heal their children.

The passage of time has rehabilitated the image of what was once the most talked-about family in the country. This is due, in part, to the Rockefeller offspring, who have aggressively defended the family name, and, in part, to some well-placed charitable donations. Baptists, especially the evangelical sort, believe water symbolizes redemption, but for the Rockefellers, it has been achieved through something more tangible: philanthropy.

Public perception doesn't always match reality, of course. Behind the arch public persona, John D. Rockefeller was a cheerful God-fearing man who enthusiastically donated to charity from the time he was a cash-strapped teenager. He was a perfectionist who championed progressive causes, including the fight for abolition, and he was passionate about medical innovation and scientific research. He cared little for public relations and shunned the spotlight, preferring the solitude of a rocking chair, and gave few on-the-record interviews. Whatever his contemporaries thought of him, it was probably wrong. What *we* think of him may be wrong, too.

John D. Rockefeller was just twenty-four years old when he left the clutches of his father, the quack, and entered the oil business as an amateur refiner. He worked initially as a middleman, transporting the precious liquid from the wells of western Pennsylvania to an expanding urban populace at the height of the Civil War. (He avoided the battlefield, hiring a substitute to take his place.) John edged out his partners to take sole control of Cleveland's largest oil refinery. Soon the company was exporting vast quantities of oil across the United States and Europe at a hefty profit. Buoyed by backroom deals with barons of the burgeoning railroad industry, Rockefeller was able to ship his product for far less than his competitors could. In short order, the trim, open-faced, blue-eyed son of a charlatan would become the richest man in America.

By the turn of the twentieth century, his company, Standard Oil, controlled 90 percent of the market, forcing rivals out of business by undercutting them. The world was introduced to the first modern monopoly, but few knew that Christianity figured prominently in Rockefeller's hostile takeovers. Rockefeller believed he was preordained to make money; it was also his God-given duty to give it away. As he waded into middle age, Rockefeller gradually extracted himself from the day-to-day operations of his company to focus on philanthropy. While Alexander Fleming was toiling in that basement in Boulogne, and Gerhard Domagk was wandering around a Ukrainian forest, John D. Rockefeller was dreaming up ways to rid himself of his vast fortune.

He was approached about starting a university. As Ron Chernow writes in his biography *Titan*, one of Rockefeller's philanthropic advisors read *The Principles and Practice of Medicine*—a doorstop of a book written by William Osler, the premier physician of the day—and had an idea: the United States should have a world-class research institute to rival the Robert Koch Institute in Berlin and the Pasteur Institute in Paris.

Although common overseas, this type of charitable giving was virtually nonexistent in the United States. Most donations in this area were targeted to specific universities or famed researchers and their laboratories, not to brand-new facilities. But Rockefeller prided himself on being a trailblazer and agreed to endow the Rockefeller Institute for Medical Research in the summer of 1901. He committed $200,000 over ten years—a large but temporary investment—to encourage medical innovation, ingenuity, and productivity. With the commitment, Rockefeller included one unique request: finances would be managed by scientists, not trustees or administrators.

US philanthropist and industrialist John D. Rockefeller circa 1897

Researchers were wooed to the institute, a scientific fantasia located in two brownstones on Lexington Avenue on the Upper East Side of Manhattan. In 1903 the university bought thirteen acres of farmland on the East River between Sixty-Fourth and Sixty-Eighth Streets, where it remains today. The Rockefeller Institute was an instant success. When a meningitis outbreak hit Manhattan in 1904, its researchers discovered the treatment that saved thousands of lives. Its proud patron pumped millions more into its coffers, an exuberant investment in medicine that perhaps represented a way to make amends for William Rockefeller's pseudoscience sham. As the tiny university's prestige and eminence grew—its investigators were the first to culture the causative agent of syphilis and unraveled how Staph infections mutate—its benefactor remained in the shadows, allowing others to take credit for the scientific progress he supported so vigorously. Although it was just a few miles from his New York City home, John D. Rockefeller visited his East River campus just once.

CHAPTER 22

Lysin

ALEX CHAPMAN* HAS AN UNUSUAL JOB. On a typical morning, after he drops his kids off at elementary school on the Upper East Side of Manhattan, he passes by Rockefeller University, dons a white coat, and turns his attention to his expanding collection of a material that most of us prefer not to think about. The forty-year-old professor has been doing this for years, approaching patients young and old, and asking gently if they'd consider donating some poop. If the answer is yes, and it often is, he or one of his assistants gathers the excrement in a special container that allows for safe transport to a massive freezer located next to Tom Walsh's laboratory. There, on the fourth floor of the hospital, buried deep within the bowels of the medical center, is the most valuable collection of feces in the world.

Chapman serves as principal investigator of a five-year NIH-funded study examining the bacteria that live in the intestines of patients, such as Remy and Donny, who have leukemia or have received a stem cell transplant. The government is betting that he can figure out why some of them get superbug infections—and, more importantly, what can be done about it. What was it about Donny's paper cut? Why was Remy the

---

* An alias is provided here.

one to acquire the bizarre fungus? Chapman is trying to find answers, one bowel movement at a time.

So far, the gamble appears to be a good one. His team uses petri dishes laced with antibiotics to identify dangerous bacteria from human excrement, looking to see if those same microbes are also causing bloodstream infections. It's difficult work—toiling in feces isn't as glamorous as it sounds—and there's no guarantee that it will produce meaningful results. But there's a growing consensus in academia that Chapman's onto something. *Something big.* He received a Young Investigator Award from the American Society of Clinical Investigation, and he is considered one of the country's most promising clinical trialists. He also works closely with scientists from Rockefeller University.

Advances in genetic sequencing have helped Chapman's group understand the diverse collection of microbes living within us. Soon doctors will be able to give their patients a printout that reveals the makeup of their bacteria—and which diseases those microorganisms might cause. A patient's diet could predispose her to Alzheimer's, while someone else's morning commute might expose him to a bug that prevents diabetes. One of the hundred trillion bacteria living in the typical human body could have a propensity to mutate into a superbug. Then what?

I bumped into Chapman in the hallway of NewYork-Presbyterian one afternoon after I had finished my daily rounds, a few weeks after the hotline meeting in New Jersey. "Just the guy I was looking for," he said. Chapman, an MD and infectious diseases specialist, was holding a stack of papers and had a steely look in his eye. "Got something for you."

"What's up?" He and I were occasionally mistaken for one another: two fortyish white guys with short brown hair who studied superbugs. I routinely asked him for medical advice, and he had once entrusted the care of one of his family members to me. Chapman provided the professional camaraderie I had been searching for in medicine; we were teammates.

"New study," he said. "I'm thinking about doing it, and I want you to be a part of it." He handed me a sheet of paper. I could make sense of most of it—it was a study looking at the treatment of staph infections—but one part was unclear. "What's CF-301?"

His face lit up. "Lysin."

"Doesn't ring a bell." Chapman and I were both mentored by Tom Walsh, but we kept our research projects separate. As junior faculty members, we were trying to demonstrate independence, and that rarely meant collaborating on trials with peers. I knew all about his work, but I hadn't contributed to it. We were teammates, certainly, but we played different positions. "Lice-In?"

"It's incredible," he said. "Check it out." He pulled out a pen and started to draw a picture of a bacterium next to a small molecule that looked like a folded protein. Chapman was a gifted researcher, but he was an even more talented instructor. I often popped into his office just to hear what he was reading and thinking about. Like Tom, he could translate head-scratching data into digestible sound bites, and he had a gift for drawing complicated structures and mechanisms—things such as efflux pumps and porin mutations, the tiny holes, or pores, that bacteria make to withstand antibiotics.

He had become one of my favorite people in the hospital, a buoyant guy whom I made a point of grabbing a beer with whenever we were at a conference together. (We somehow never had time to hang out in our own city but managed to do so away from home.) We'd compare notes, discuss upcoming talks, exciting new drugs, and gossip. He was someone that other hospitals and universities were eager to recruit. "I'm thinking about this lysin trial," he said, "and I want your help. If I do it, I'd want your help."

"Another trial?" At last count, Alex was managing at least a dozen.

"Thought you could help."

"How do you spell it?" I asked, trying to follow his drawing. "L-y-s—?"

"It's pretty cool stuff." He sketched out a few structures. At the top, a protein was entering bacteria. As he scribbled, my mind drifted, as it often did when someone took their eyes off of me. When I bumped into Chapman, I had just finished seeing a patient with MRSA, and to my dismay, the man wasn't responding to treatment. I was nervous, and my patient knew it. I made a mental note to discuss the case with Tom.

"I'm listening," I said to Chapman as he continued to draw. I studied the picture and scratched my chin. "It just causes the bacteria to explode?" I made my right hand into a fist and then extended my fingers. "Just like that?"

"Something like that, yeah."

"Why have I never heard of it?"

"New to me, too."

Sensational claims of treatments for superbugs were not uncommon— the most recent I'd come across was a tenth-century eye salve remedy from the Old English medical text *Bald's Leechbook*—but the hype was always short-lived. We weren't just going to stumble on a cure the way Alexander Fleming did. If lysin actually worked, I would've heard about it. Or so I thought. "Have you talked to Tom?"

"I have," Chapman said. "He's in."

"Let me do a little digging," I said. "But yeah, count me in, too." We cemented the deal with a handshake. The study would be sponsored by ContraFect Corporation, a biotechnology firm. These companies use the biological processes of microbes to create new medicines and technologies to treat diseases in humans. "They're based in Yonkers?" I asked, somewhat amused. "They have biotech?" I passed by there every morning on the way into Manhattan; I didn't realize Yonkers was becoming a hub of innovation.

Chapman handed me his drawing. "If you want know more about it . . . this lysin stuff was discovered just down the street."

"Is that right?"

"All the work was done at Rockefeller." My hospital had formed a tri-institution alliance with neighboring institutes—Rockefeller University and Memorial Sloan Kettering—to train physician-scientists to make these kinds of discoveries. Every year, more than five hundred people apply for a handful of research positions, all fully funded by the NIH. They are among the most prestigious and competitive spots in the country.

Chapman's phone buzzed, and we both paused to review our

in-boxes. Email had become so unmanageable for many of my colleagues that they'd stopped using it as a means of communication. Now
we were receiving text messages every few minutes. At the end of every day,
I make a point of deleting all of my texts, just so I won't feel overwhelmed
the next morning. I picked up the habit after reading a *GQ* magazine profile of Kim Kardashian in which she was described as "frighteningly
organized" and mentioned the nightly need to expunge texts.

"Not entirely sure I'm gonna do this study," Chapman said after he
responded to the new message, "but it's worth checking out." He leaned
back on his heels and let out a deep breath, indicating that he had somewhere to be.

"More poop?" I asked, pointing at his phone. He nodded, and we
shook hands, off to consent more patients for our respective trials.
"Make sure you wear gloves."

CHAPTER 23

Breakthrough

AFTER THE IMPROMPTU MEETING in the hallway, I returned to my office and tumbled down the lysin rabbit hole. *How had I missed this?* I also tracked down the man responsible for developing it, an immunologist named Vincent Fischetti, who has been doing research at Rockefeller University for nearly fifty years. He invited me to discuss the work at his laboratory, where I could see his experiments for myself.

A few days later, on a brisk, foggy morning, I walked down York Avenue, past the Rockefeller tennis courts and the Philosopher's Garden fountains, imagining the campus as it had been a century earlier, before skyscrapers and iPhones and Uber—before the richest man in America first set foot on it: cows grazing on a treeless pasture, barges floating up and down the East River, the Queensborough Bridge being built off in the distance. Steam—steam rising everywhere, from the river and the slaughterhouses and the soot-stained chimneys surrounding the muddy farmland. A time before antibiotics, when the average life expectancy for a guy like me was forty-seven. I was nearly there.

The entrance to Fischetti's laboratory building displays a large poster featuring black-and-white photographs of elderly white men, titled *Nobel Laureates of Bronk Laboratory*: Gerald M. Edelman, Ralph M. Steinman, George Palade, Günter Blobel, Fritz Lipmann, and Christian de

Duve. Fischetti works on the eighth floor of Bronk, at the end of a long hallway overstuffed with expensive equipment and hirsute graduate students. His office is slightly removed from their laboratory benches, across the hall from all the lysin work.

When I knocked on his door, he sprang up. The professor is in his seventies, but he appears much younger. Fischetti has a perpetual tan, powder-white hair, even whiter teeth. "Thank you for meeting me," I said as I removed my white coat and took a seat next to a framed picture of exploding bacteria. I explained my role as a physician-researcher at Cornell and mentioned that I might be an investigator on his upcoming lysin trial.

He pointed at the image I was looking at and smiled. "That's lysin in action. Incredible, right?" Like Tom Walsh, Vince Fischetti has the ardor and enthusiasm of a man half his age.

I nodded at the mini-explosion and pulled out a notepad. "I'm going to be consenting patients," I said. "So I figured I should know how this stuff works. I've read some of your papers. Wild stuff."

I discovered that Fischetti has spent decades trying to understand the changes that take place when bacteria interact with human cells. His Rockefeller team tries to identify and interfere with these microscopic events, using enzymes derived from bacteria-killing viruses (known as bacteriophages) to prevent and treat infections. "It sounds risky," I said. I scanned his office and silenced my phone. "Injecting people with bacteria-killing viruses? I'm worried that our IRB will never approve it. I'm not sure I'd want it."

He shook his head. "That's not quite right," he said. "We've removed the virus and purified the protein. Let's back up." Fischetti started from the beginning. Lysins are enzymes that have evolved over a billion years to degrade bacterial cell walls, he told me. They're highly specific—there's a lysin for nearly every bacterium—and they're not subject to bacterial resistance. Unlike antibiotics, they don't weaken over time.

Fischetti has been trying to harness the power of lysin for years—he first purified the protein when Nixon was president—but his big breakthrough came in 2001. That year, he published a landmark study in the journal *Proceedings of the National Academy of Sciences of the United States* showing that lysin could be used to treat infected animals and cure them. Lysin had been studied extensively in test tubes, but no one had shown that it was effective in animals or humans. "No one thought it would really work," he said. Fischetti was the first to show that a single dose of lysin could protect mice that were exposed to ten million strep bacteria—the same ones that had devoured Fleming's soldiers on the battlefield. A second experiment, similar to the one Gerhard Domagk had performed, revealed that when lysin was given to infected mice, strep disappeared two hours later. "Let me add a little lysin to see if it kills bacteria in mice," Fischetti recounted, moving his hand like a chef garnishing a dish. "And it worked!"

It was as if the bacteria had been wiped away with penicillin or sulfanilamide, and it suggested that lysin might one day complement or replace antibiotics altogether. But there was a problem. "It took a long time to get anyone from industry to invest in our work," Fischetti said with a touch of amusement. "We were focused on targeted killing, but Big Pharma doesn't want to hear that. They want broad spectrum!" Companies passed on lysin, figuring it wouldn't turn a profit. But Fischetti was undeterred, and he had the resources at Rockefeller to continue his work. He knew he had a story to tell, and, like Fleming, he just had to write it.

In a string of papers in top journals, he continued to show the potential of lysin therapy as an alternative to antibiotics. He began purifying and cloning and analyzing all sorts of lysins, and eventually ContraFect acquired the rights. The small company made it a part of its research portfolio, rebranding lysin as CF-301, and in 2015 the molecule received fast-track designation by the FDA. In 2016, fifteen years after Fischetti's initial experiments in mice, ContraFect announced that phase 1 trials in humans had been a success. "We have successfully concluded the first in-human trial of CF-301, a first-in-class lysin drug candidate," Dr. Steven C. Gilman, the company's CEO, announced. "We are now expeditiously moving towards the next phase of development of CF-301, which will be in patients with *Staph aureus* bacteremia."

That's where Dr. Chapman and I come in. Lysin had been tested for safety in healthy volunteers and performed beautifully, but it hadn't been evaluated in patients with staph infections—the ones who desperately needed treatment—and it was time for doctors to recruit them. Obtaining patients' consent for any clinical trial is serious business, but it becomes a high-wire act when the patients are really sick; roughly a quarter of men and women with *Staph aureus* in the blood will die from their infection. I came to Fischetti because I wanted to understand the science inside and out before I approached a vulnerable patient with a consent form. It would be impossible to get informed consent unless I was fully informed.

Fischetti summarized his remarkable career for me, hopscotching across experiments that laid the groundwork for the upcoming staph trial. I kept shaking my head. Why hadn't I heard about any of this? No one had mentioned lysin in college or med school or residency or fellowship training. As he spoke, my eyes wandered around his cluttered office. This was a man ahead of his time. He had performed these groundbreaking experiments largely in anonymity—he was a giant in his field, but somewhat unknown outside of it—and he was finally about to get his due. Fischetti was starting to recruit speakers for an international lysin conference, to be hosted at Rockefeller University, where his life's work would be celebrated by some of the world's top scientists. I made a note to be there for it.

I marveled at lysin's origin story: John D. Rockefeller's fascination with medicine led to a top-flight research center, with geniuses like Fischetti funded extravagantly by an endowment that would ensure scientific progress for generations. *Redemption through philanthropy.* And now I might get to be a part of it: I would help find patients who were willing to try lysin.

The economics of Rockefeller University threw into relief the dire straits my own field was in. Infectious diseases specialists have become a dying breed in some parts of the country, cast aside by modern medicine. Most doctors are now compensated based on the types (and cost) of procedures they perform, and infectious diseases doctors don't really perform procedures. Ours is a cognitive specialty, providing expert consultation, and reimbursement schemes haven't figured out how to keep up with the tremendous demands of the work. The field is experiencing a brain drain, and every year, it gets a bit worse. Specialists still flock to big cities on the coasts, but the middle of the country has been hit hard by the changing economics of medicine. Young doctors are less interested in infectious diseases than their predecessors were, and this presents a problem: once lysin is approved, there need to be specialists who know how to use it.

One problem is that training grants have become too restrictive. Budding physician-scientists are asked to choose between the laboratory and patient-based research, which creates gaps in areas such as public health, epidemiology (the study of the distribution and determinants of disease), and the kind of translational work that's needed to bring Fischetti's discovery to patients. Not many people can do it all the way Tom Walsh can, but we've stopped giving young investigators the opportunity to try.

Over the next hour, Fischetti brought me up to speed on his current work involving vaccines and antibodies. "Take a look at this," he said, handing me a manuscript. His team had just published a report proving that lysins could be attached to human antibodies—Y-shaped proteins released by the immune system to neutralize pathogens—producing "lysibodies" capable of protecting their hosts from a variety of infections, including MRSA. "We're also developing a lysin spray for burn patients," he continued, to protect them from the superbug *Acinetobacter* as well as one targeting *C. diff*, the highly contagious diarrheal infection that nearly killed George, the World War II veteran in my trial.

He motioned me over to his computer and pulled up a video. "Here's a bacterium," he said. "Think of it as a water balloon. Lysin punches holes in the cell wall, just like punching holes in the balloon." My eyes darted across the screen. "You don't want to punch too many holes," he warned as he showed me a video of a lysin destroying the bacterial blob. "Too many holes cause inflammation, and that's a problem." He looked up from the screen. "The dose you'll be using in the trial will just punch a few holes."

In one afternoon, Fischetti convinced me that the notion of broad-spectrum bacterial warfare has become impractical. New dangers are appearing too quickly, and we can no longer afford to wait for the next miracle drug to wipe them out. The immunologist's work sends a clear message: we need to attack bacteria one by one. It will take years to bring the treatment to patients—every lysin will require its own approval by the FDA—but it could be worth it. "I really want to try lysin with prosthetic joint infections," he said as I was getting up to leave. "I just need someone . . ." His voice trailed off, and he scratched his cheek. "Someone willing to take a chance. Someone who knows how to do animal studies who's willing to take a risk."

"I think I might know someone." Fischetti and Tom Walsh were ideal collaborators. Connecting them was probably the most useful thing I'd do all day. "I'm seeing Tom this evening," I said. "We'll draw up a protocol. This sounds like a great project."

"Based on everything we've seen in animals," Fischetti said, "this has got to work. It has to!" He paused for a minute and stared at a molecular diagram. "Unless . . . unless something happens that we haven't predicted."

I frowned. "Like what?"

Lysin looked good in the laboratory, but there was no guarantee it would work in humans. That's why we needed to perform the trial, so we could address a host of crucial questions: Would lysin cause overwhelming inflammation? Or an allergic reaction? What if we punched too many holes in that water balloon? We stared at each other for an uncomfortable moment, then my eyes settled on an issue of the scientific journal *Nature* next to his desk. Fischetti followed my line of sight and pointed at the cover. "Oh, that's anthrax," he said. "I haven't told you about anthrax."

CHAPTER 24

Anthrax

THREE WEEKS AFTER the World Trade Center crumbled, a man in Florida woke up feeling unwell. After a mostly sleepless night, Robert Stevens had a fever and chills, and before long, he was vomiting. His wife, Maureen, brought him to an emergency room near their home in West Palm Beach, where she mentioned to doctors that Robert, a sixty-three-year-old photo editor, had been dealing with vague symptoms for several days, including muscle aches and malaise. She couldn't put her finger on it, but it was different from the other times he'd been sick. Robert simply wasn't himself.

Her husband was disoriented and unable to provide the doctors with much information, so Maureen did most of the talking. She told the emergency room staff that Robert was a typical Floridian, passionate about fishing and gardening, and that other than some high blood pressure and mild heart disease, he was in good health. His only medications were a beta-blocker and a baby aspirin. Robert worked in Boca Raton at American Media, publisher of the *National Enquirer* and other tabloids, and he spent most of the workday reviewing photographs submitted by mail or over the internet.

Doctors weren't entirely sure what to make of the case. The patient's white blood cell count was normal, which argued against infection, but the combination of fever and confusion was enough to suspect bacterial meningitis, and the doctors started him on Mississippi Mud and another antibiotic called cefotaxime. They told Maureen that Robert needed a CAT scan and a lumbar puncture (spinal tap) to confirm the diagnosis. Things went downhill from there.

Just a few hours after arriving in the ER, Robert suffered a grand mal seizure and was placed on a ventilator. The lumbar puncture was wildly abnormal—the normally clear spinal fluid was cloudy, indicating a massive infection, and it was teeming with bacteria. But they didn't look like typical species one sees in spinal fluid. They were long and slender, and they were also swimming in his blood. Over the next twenty-four hours, Robert's condition worsened. His temperature rose to 104 degrees, and his kidneys shut down. On the third day of hospitalization, Robert's blood pressure plummeted, and he went into cardiac arrest. Resuscitation was attempted, but it failed, and he was pronounced dead. By then, the Florida Department of Health had confirmed publicly what those caring for Robert already knew: he was killed by anthrax.

It was the first such death in the United States in twenty-five years, and it was unlike anything epidemiologists had seen before. For the past century, most cases were due to occupational exposure to contaminated animal hides—anthrax, a rod-shaped bacterium known more formally as *Bacillus anthracis*, occasionally infects people who work with goat hair—but Robert Stevens sat in an office all day sorting pictures. Yes, he had taken a trip to North Carolina before his symptoms began, but nothing had been out of the ordinary; there was no hunting or visits to farms or petting zoos. He certainly didn't touch any goats.

The Palm Beach County Health Department, the Florida State Department of Health, and the CDC launched a wide-ranging investigation in conjunction with the FBI as epidemiologists descended on Robert's home and office. One sample taken from his office building tested positive for anthrax, and it was soon found in the nose of Robert's seventy-three-year-old coworker. The Health Department began offering prophylactic antibiotic treatment to anyone who might have been in the building for the past sixty days. The cone of exposure was enormous, and more than one thousand people underwent nasal swabs in the weeks after Robert's death. Complicating the investigation, people who were exposed to anthrax reacted differently—some could inhale it and feel fine, while others knew immediately something was wrong and died in a matter of days. It was hard to know who was truly at risk.

One week after Robert Stevens's cardiac arrest, the New York City Department of Health alerted the CDC that it was in contact with a person harboring a strange rash. It was a black lesion, known as an eschar, which is sometimes mistaken for a brown recluse spider bite. The eschar isn't tender, but it swells and weeps fluid, and it's caused by only one thing: anthrax. Shortly after that, the CDC was notified of another dark rash spreading rapidly on the left arm of an infant. Anthrax had long been considered a potential weapon for bioterrorists, and now it had arrived in the most densely populated city in the country.

One of the eschars was on the chest of a thirty-eight-year-old woman who worked for NBC TV. Days earlier, she had handled a suspicious letter at her office. The letter contained a powder, and that powder contained anthrax. The FBI announced that the NBC envelope and another that was sent to the *New York Times* were both postmarked from Saint Petersburg, Florida, and had similar handwriting. A few days later, Senate Majority Leader Tom Daschle announced that anthrax had been found in his office, too. Two weeks after that, two postal workers from the Brentwood facility in Washington, DC, were confirmed to have died from anthrax inhalation. As investigators homed in on a suspect, the outbreak expanded. More than ten thousand postal workers had potentially been exposed to anthrax, and experts were needed to figure out who required immediate treatment.

Tom Walsh was in DC at the time and dropped what he was doing to examine postal workers and their families, carefully determining whom to treat with ciprofloxacin. "A boy came into the clinic and vomited everywhere," Tom recalled. "The guy doing triage looked at me and then at the vomit and said, 'This one's yours, Walsh.'" He and other first responders ultimately gave prophylactic antibiotics to twenty-five hundred people. At the time, anthrax was considered to be 80 percent to 90 percent fatal even with treatment, but Tom's team helped to cut that number in half.

Five people died during the anthrax attacks of September and October 2001, which was likely the work of Bruce Ivins, a disgruntled government scientist who worked on anthrax vaccines at the US Army's biodefense laboratory at Fort Detrick, Maryland. (Ivins committed suicide before the investigation, led by FBI director Robert Mueller, was complete.) Ten years later, in 2011, Maureen Stevens was awarded $2.5 million by the US government after arguing successfully in court that it had not adequately secured its supply of the dangerous pathogen. But that belated settlement was not the end of the story.

In 2016, Russian health officials responded to an anthrax outbreak in Siberia, where dozens were hospitalized and one child died. Families were airlifted out of remote villages while investigators tried to piece it all together. Epidemiologists ultimately settled on an odd theory: decades earlier, anthrax had killed reindeer, and the carcasses had remained under a layer of permafrost until a heat wave hit the Russian peninsula, bringing the deadly spores to the surface. If the hypothesis was correct, it meant humans would be at risk every time temperatures spiked. Anthrax wasn't going away; if anything, it was coming back.

While Tom Walsh was triaging postal workers, Vincent Fischetti was drawing up experiments. His team at Rockefeller developed a lysin to detect and destroy anthrax bacteria, and the discovery landed him on the cover of *Nature*, the journal I was now staring at in his office. "A new drug could foil bioterrorists' attempts to engineer antibiotic-resistant anthrax," it reported about his work. "The drug could also make a quick hand-held detector for checking contaminated sites."

Fischetti pulled up a video on his computer and waved me over. "This is anthrax," he said, pointing to a grey rectangle on the screen. "And this is lysin." Two clicks later, the bacteria had burst like a water balloon. As the video played, I scanned the old *Nature* article on my phone: "The team plan to start clinical trials on animals within weeks. The drug could be ready to stockpile within three years, predicts Fischetti."

"Nice," I said. "Is this available for patients? Or in clinical trials?"

Fischetti shook his head.

"Really?" The article was more than fifteen years old. Surely some progress had been made? Fischetti played the video again and again and again.

CHAPTER 25

Delivery

I IMAGINED THE conversations I would have with my colleagues, convincing them that there was a safe way to make bacteria explode. I made the short trek from Fischetti's office over to see Tom, considering the various ways to describe the lysin trial. The balloon analogy seemed like a reasonable place to start. *We'll punch holes*, I thought as I knocked on Tom's door, *but not too many holes.*

Tom always knows when it's me. I have a characteristic rap—two hard knocks in quick succession—and when he hears it, he shouts, "Come in, Matt!" I took off my white coat and let myself in. Tom spends most of his time doing research and occasionally misses the immediacy of seeing patients. So when I have a free moment, I involve him in my cases, especially the challenging ones. "Another *Candida auris*," I said as we sat at the conference table in his office. "I'm giving micafungin for now." We both knew that the antifungal drug would soon stop working, and when it did, I was prepared to give the new antifungal drug, the one made by Scynexis.

I wanted to review my patients before we got to lysin. I pulled out my list. "Let's see, I've got a patient with cellulitis. Might be a candidate for dalba; the drug should *finally* arrive today . . . oh . . . I have a firefighter who came in with shortness of breath; might've inhaled something. But he's got a crazy backstory." Tom raised his eyebrows. "He's NYFD, and on 9/11 his engine company was pulling people out of office buildings, bringing everyone in the World Trade Center down to the lobby. Telling them to stay put."

"Oh, no."

"But this one older woman on the twenty-first floor insisted on being carried out of the building. So my patient threw her over his shoulder and carried her three blocks north. Six minutes after they left the building, it collapsed." I paused, thinking about how others reacted to that awful moment: Donny looking for his gear, Tom Walsh gathering volunteers at the Stanhope Hotel. "After the attack, my patient developed PTSD. He blamed himself for all of the people who died in the lobby, and he was the only one from his engine company who survived."

I took a sip of water, and Tom grimaced. "Three months later, my patient gets a phone call. It's the woman. She called and said, 'You saved my life,' and offered to take him out to dinner as a thank-you. He agrees and points out that she also saved his life. When he shows up for dinner, her entire family is there to honor him. Over dinner, he hits it off with the woman's daughter—and the rest is history. They fall in love, get married, have a bunch of kids."

"No!"

"Yes!"

"That's incredible."

I scanned my list. "I also have a homeless guy crawling with bugs." I reflexively scratched my forearm thinking about him. "Plucked one out of his ear."

"Really?" Tom's face lit up. "Do you have any of them? Shall we take a look?"

"The bugs? I just sent one to the micro lab."

He grabbed his white coat. "Let's go." Moments later, we were staring into a microscope. A dead critter was lying on its back and missing one of its tiny legs. "What do you see?" Tom asked. He put his hands to his face, quietly rooting me on, hoping I knew the answer. "Look closely."

I peered back in and shook my head. "Not a tick. Not a parasite. I'm not quite sure."

He saw an opportunity. "Body louse," he said. "Are you familiar?"

A few of the technicians wandered over. "Vaguely," I replied.

Tom smiled—an impromptu lecture was about to begin. "In his book *Rats, Lice and History*," Tom said to the growing peanut gallery, "Hans Zinsser describes the devastating impact of typhus fever throughout history, including on US troops in the First World War." I took out a pen as he swiveled back and forth, so everyone could hear his soft voice. "Zinsser served in the Great War as a medical officer," he went on, "and was a graduate of Columbia University. He is eponymously remembered by the Brill-Zinsser disease."

When Tom launches into one of his impromptu didactic sessions, it feels like liftoff, bringing those around him into an unexpected and unexplored space. A hush falls over the room, and everyone inches forward to hear this soft-spoken man explain the unexplainable or pry open the lock on a medical mystery. Occasionally there is turbulence—I often force him to slow down, to explain what is so clear to him but inaccessible to others—but it's always worth the journey.

I scribbled the name *Zinsser*. Then I felt my phone buzz. There was a text message informing me that one of my patients, a man from the Bronx named Steven, who had the inflammatory connective tissue disease lupus and an opiate addiction, was in the emergency room with a low-grade fever. I'd cared for him a half dozen times over the past year, usually after he ran out of OxyContin, and, as a courtesy, the ER had informed me of his arrival. "To be continued," I said to Tom, patting him on the shoulder.

When I entered Steven's room, I found him surrounded by young doctors. He looked malnourished and exhausted; his cheeks were sunken, and his neck veins were pulsating. He was not the vibrant guy I remembered. His arms now looked like matchsticks, and his chest was lopsided. Steven had been given three antibiotics, and the medical team wanted to know if he needed a fourth. Over the past year, I had become his de facto doctor because he kept missing appointments with primary care specialists, occasionally refilling medications for him when he ran out. I examined Steven and removed the stethoscope from my ears. "No more antibiotics," I said to the team. "I don't think he's infected."

It was a position I increasingly found myself in. Rather than dispensing antibiotics, I was functioning as a steward, trying to withhold them. Overuse was promoting the development of superbugs—most doctors know that—but it's hard to resist when you see a patient with a fever and plummeting blood pressure. "The fever is from lupus," I added. Withholding antibiotics from patients like Steven is just a drop in the bucket, of course. The spread of superbugs is driven largely by improper animal husbandry, poor sanitation, weak infection-control policies, and overcrowding—that's why they often pop up in New Delhi and New York. By saying no occasionally, I get to feel like I'm playing my small part. "No antibiotics."

The dalba trial was going to turn this approach on its head. Instead of saying no, I would be trying to convince patients to say *yes*. Yes to an antibiotic that I had never given to anyone. Yes to a new drug that might not work. I was going to ask sick patients and their doctors to trust me. The first dalba shipment arrived several hours after I saw Steven in the emergency room, and when it did, I didn't think of the patients it might help. I didn't think of the data I would gather or the papers I would eventually write. I thought of Omniflox, the drug that was recalled shortly after approval because it killed people. When I held the first shipment of dalba in my hands, I had to stifle a deep wave of apprehension. After a half year observing patients such as Ruth and George and Erwin and Donny, it was finally time to get started.

PART 5

Toward a Cure

CHAPTER 26

Meghan

FOR THE FIRST six months of the dalba study—the preperiod—I simply watched, documenting what happened to hospitalized patients with skin infections who received routine medical care. Most stayed in the hospital for about four days—some much longer than that—while receiving Mississippi Mud or another intravenous antibiotic. Some, like George, did just fine. But there were complications, of course, including accidental falls, blood clots, brain bleeds, and hospital-acquired infections, and my job was to keep track of what happened. After discharge from the hospital, these patients returned to my office a few weeks later, so I could see if their infections had resolved. Some felt worse, some felt better, and all of them wanted to know what could be done to prevent it from happening again.

In those follow-up appointments, I learned that Ruth and her family had decided against a feeding tube after a palliative care specialist—an expert in life-limiting diseases and end-of-life care—explained the risks of placing a foreign body in her stomach. Anne accompanied her to my office and told me that her mother was encouraged to have frequent meals and to take small bites, and to mince her daily medications into applesauce. "It's a blessing," Anne said, "that she didn't get the tube. It wouldn't have helped, and it might have hurt." After our appointment, they planned to go shopping.

These visits provided the continuity I had been seeking in our fragmented medical system, and I found myself wishing that the volunteers were my own patients—men and women whom I would advise and treat for years. George had been reunited with his parakeets and refused to accept compensation for participating in the trial, while Erwin, the medical student, told me he planned to get his nipple pierced at the site of his bite mark to commemorate the event. He said to look him up if I ever needed an operation.

They weren't all happy endings, however. Piper Larson's stomach cancer had indeed spread to her lymph nodes, as well as to her liver and her spine. By the time I had returned to see her a few days after our initial encounter, she had already met with a hospice nurse and was planning out the last few months of life. I didn't know what to say to her, and I didn't even try to find the right words. I just squeezed her hand and thought of her son.

From this six-month observation period, I discovered just how much room there was for improvement. Patients still had questions, and some still had pain. This information was necessary to establish a baseline, to provide some understanding of how a new antibiotic might improve the status quo during the postperiod, which we'd refer to as the interventional phase.

The first person I approached about taking dalba was a woman who was dealing with an agonizing and inscrutable symptom: pruritis. Meghan Darling had been retired for just two weeks when her right leg started tingling. "I was going to bed," she said from her stretcher in the hallway of the NewYork-Presbyterian emergency room, "and all of a sudden my leg starts throbbing. Then it turned into these tiny little pinpricks that I needed to scratch. I went to brush my teeth, and I nearly fell over, but my husband caught me." Meghan limped into my hospital a day after the first shipment of dalba had arrived, searching for answers.

Darling had spent three decades in the publishing industry, editing children's books that were popular in Europe so that they would be palatable to American readers, before taking a buyout and moving from New York City to the suburbs of Westchester County. "We're planning a trip to the beach to celebrate retirement," she said, "and then this happens." She pulled up her denim pant leg to reveal her right shin. It looked like maggots had been feasting on her. I tried to mute my response, but my poker face failed me. "I know," she said, shaking her head. "Absolutely disgusting."

I crouched down to examine the leg. "How long have you been living with this?" An area of dead skin resembled the shape of Texas, with sharp points on the left edge and bottom. I took a deep breath, hoping the aroma might tip me off to the type of infection—strep can smell like butterscotch; *Pseudomonas* evokes grapes—but there was nothing. Meghan's leg was just a mass of inert skin and scar tissue.

"Too long," she said. "Far too long."

A closer look revealed that a purple crater existed where skin should have been. Open blood vessels led to tiny blood clots all over the lower leg; some areas were yellow, while others were mauve. It looked like Freddy Krueger's mottled skin. "Does it hurt?" I asked.

"Some days," she said. "Other days it just itches."

I took a step back and grabbed a pair of latex gloves. "Mind if I touch it?"

"Do your worst."

I tapped on the wound, expecting Meghan to flinch, but she didn't. I pressed deeper, hoping to elicit a reaction. No response. "Feel anything?"

She shook her head. "Nothing. Today it's numb. Last week it was itchy."

The lack of sensation made me think of leprosy, but that was an unlikely diagnosis for an editor who worked in an office building in New York. "Have you done any traveling?" I asked. "Anywhere exotic?" I had once diagnosed a woman from Brazil with leprosy, but never someone from the United States.

"Just Disney," she said. "And Epcot, if that counts." She giggled. "And Daytona Beach." Meghan spent the next ten minutes walking me through the details. During that trip to Florida, shortly after the tingling started, her discomfort worsened. "I was scratching and scratching all day and night. One morning I woke up and literally had chunks of skin under my fingernails." She held her palms upward and examined her hands. "There was blood on my fingertips."

"Did you see anyone about this? A doctor?"

"I had this harebrained idea that the seawater would cleanse the wound, so I stayed in the ocean for hours. And when that didn't work, I soaked it in the pool, hoping the chlorine would help."

"Did it?"

She looked down at the leg. "What do you think?"

I nodded. "It just looks so . . . painful."

"Eventually my husband convinced me to take Benadryl. He said the scratching was driving him nuts and begged me to do something."

"And?"

"Benadryl made it worse!"

"I'm surprised," I said. There were a variety of medical conditions that caused intense itchiness—liver and kidney problems and several cancers—but most responded to antihistamines such as Benadryl. I studied Meghan's anxious face; this was not the time to mention leprosy or malignancy.

"You and me both."

The more I listened to her, the less convinced I was that she had a skin infection that would respond favorably to dalba. I wanted to make sure we were using it in the right patients, and that meant people who had bacterial infections of the skin. But I didn't know what Meghan had—it was unlike anything I had seen during the first part of the trial. Meghan sensed my confusion. "Did you ever see the movie *Alien*?" she asked. "Because that's my leg."

We stared at the violaceous ulcer. "And you never saw anyone about it? Never talked to your doctor?"

Meghan ran a hand through her thick, grey hair and grimaced. "I withdrew," she said. "The more it grew, the more I withdrew. And then . . ." Her eyes welled up suddenly, and tears spilled over onto her cheeks. "My husband left me eleven days ago." She covered her eyes with her right hand and began to sob. Then she looked up at me and let out a deep sigh.

"I'm so sorry." I leaned in gently, as if to offer a hug, but stopped. The emergency room was a difficult place to have a moment like this. I folded my arms and took a step toward her. Then I rubbed her damp back. "I'm just so—"

"It's okay," she said. "You're allowed to hug me." Meghan tried to muster a laugh and extended her arms. "And don't be sorry. It was a long time coming."

I grabbed a handful of paper towels from above the sink next to her stretcher and offered them to her. "We have people here if you want to talk."

Meghan wiped her eyes and looked down at her leg. "I just want this to go away."

Our conversation veered away from informed consent as we meandered through the narrative of her illness. I learned that Meghan had finally seen her primary care doctor a week after her husband walked out, when the leg started oozing blood. "The doc said he was worried that my leg was weeping . . . and when he said that . . . I just lost it. Tears from my eyes and tears from my leg. Weeping everywhere."

"You've been through a lot," I said. "More than anyone should have to."

"It just got to be too much, you know?"

Beneath her sadness, the recurring thread through her story was embarrassment—profound embarrassment that a part of her body could fail her so. It's why she avoided physicians and pulled back from the world. I put a hand on her shoulder and gave a squeeze. "We will get you through this," I said firmly. For a moment, the room fell silent. Human connection had the power to drown out the noise, to transform a conversation into something transcendent.

Meghan bit her lip. "I hope so."

I pulled out the consent form and explained the details of the study. She was willing to participate, but there was a complicating factor: Meghan had been brought to my attention because an emergency room doctor had admitted her to the hospital for treatment of a bacterial skin infection. After examining her, however, I still wasn't so sure that was the proper diagnosis. Her condition looked more like something called pyoderma gangrenosum, a skin disease that affects one in one hundred thousand people. I usually see one or two cases a year, but they were never this bad, and they rarely caused itching. Dalba might help—staph and strep had been found in some ulcers caused by pyoderma—but it wouldn't fully eradicate the disease. Doctors weren't sure how to cure it, but some had found success with steroids.

"Be honest with me," Meghan said while dabbing her eyes. "Have you ever seen something this—what's the word?—nasty?"

"Sure," I said. Her skin recalled a case I'd handled in the burn unit six months earlier, when I'd been asked to see a Brazilian woman as part of the hospital's medical ethics committee. That patient had doused herself in gasoline and struck a match after finding out that her husband had been unfaithful. Eighty percent of her body was burned. The surgical team had removed as much of the dead skin as it could without killing her and finally asked our committee about palliative options.

"If it makes you feel better, I can definitely say I've seen worse."

"I'm not sure it does." She thought for a moment. "Maybe it does."

"The thing is," I said, "the more I look at your leg, the less I'm convinced it's an infection."

"Well," she said, "if you want me for the trial, I'll do it. If not, no problem." She took out her phone and tapped out a text message.

I studied my consent form. "So tell me about your job. What does it mean to Americanize a book?" I asked.

"I could talk your ear off about that."

"Go for it," I said. "I got time."

She motioned me close to her, so passersby couldn't hear. "Here's the thing: children's books in Britain can be super racist."

"Oh?"

"Like, unforgivably racist. They'll have drawings of people from other countries, almost exclusively colonial countries, with oversized features—huge lips and noses—speaking broken English."

"And you fix that?"

"Bingo." I could see a flicker of the old Meghan, the successful editor, the woman before the illness. "And some stuff is just strange. One German book I worked on claimed ballerinas put their hair up in a bun so they don't get dizzy. Nope! It's so the hair doesn't get in their eyes. Everyone knows that. I fix little stuff like that, too." She became more animated as she spoke. But after a few minutes, I noticed that she was staring again at her leg, pausing for long stretches, and losing her place in her stories. I could see that she was trying to resist the urge to scratch. "Please . . . please make this go away," she said as she dug her nails into the ulcer. Tiny drops of blood percolated from the leg after she withdrew her hand.

"I'll do my best. I'm going to call your doctor shortly."

Meghan motioned me toward her again. "You want to know how my husband left me?"

I wasn't sure there was a right answer. "Do I?"

"Email."

"No!"

"One paragraph."

I exhaled slowly, marveling at the ways patients let me into their lives. On many days, including this one, I thought to myself, *I don't deserve this. I haven't earned it.* The white coat invited confidence. I learned things about people that they would never share with their closest friends and family. I had been raised Catholic in a Florida suburb, periodically going to confession, and now I felt like I was on the other side of the dividing wall.

"Left me for his high school girlfriend. Bumped into her at Knicks game."

I could tell it was something she hadn't shared with many people, if any. The emotional wound was still fresh. "The Knicks?"

"And you know what? That email was full of typos." Meghan shook her head. "The guy needs an editor!"

CHAPTER 27

Mantra

AFTER I SAID GOOD-BYE to Meghan, I headed up to my office to review her file. A surgeon had determined that she might benefit from a special boot to relieve pressure from her leg, but the fitting wouldn't occur until her infection had been treated. Then I walked over to meet the hospitalist assigned to her care. He was an infectious diseases specialist who, like me, cared primarily for patients on the general medicine service. "What do you think?" he asked when I stepped into his office.

"What do *you* think?" I replied. "Interesting case."

"I don't think it's cellulitis," he said, contradicting the opinion of both the emergency room doctor and the surgeon. "She needs wound care and a dermatology consult, not an antibiotic. Possibly steroids."

"Agreed," I said. "Although there could be a bacterial superinfection."

"Sure, I suppose. But I don't think she's a good candidate for dalba. Looks like pyoderma to me." I returned to the ER to deliver the news to Meghan. We were going to help her, but it wouldn't be with antibiotics, and it wouldn't be with dalba. I was eager to give her my new drug, but she was the wrong patient. I apologized for wasting her time.

Half an hour later, I returned to my empty office. I nudged the mouse on my desk, and the computer screen flipped on, illuminating the dark room. I pulled up Beethoven's Moonlight Sonata on YouTube and

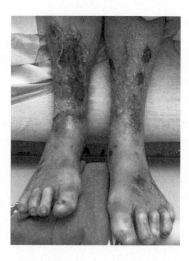

*A research subject with leg infections that
did not improve with oral antibiotics*

removed my white coat and stethoscope. The dalba study would have to wait. More patients were coming, certainly, but no one qualified yet for my new antibiotic. As the music wafted into the room, I opened a spreadsheet with data from my trial. It had reams of information about George and Ruth and Donny and scores of other patients. Below it were notes about patients I was following from afar—children like Remy—as well as reminders to read up on their diseases. Next to Donny, I had written "paper cut what?" and pasted a link to a paper I'd been meaning to read—one that might provide some insight about his bizarre condition. In that moment, the to-do list felt overwhelming.

I turned off the music. In some ways, I was becoming more like my mentor, but this was not one of them. Classical music mostly remained inaccessible to me. Tom had exposed me to it for years, but I still didn't appreciate it the way he did. The sonata sounded like a death march, and it was bumming me out. I put on "Take It to the Limit" by the Eagles.

I occasionally found myself buckling under the weight of my work, struggling to balance the demands of a clinical trial with daily patient care and a busy home life. My wife is a transplant nephrologist across town at the Columbia University Irving Medical Center, and we have two young children who like to begin the day long before we do. On some days—many days—the stress of it all can bring me down; it feels like I'm just running out the clock, competing in a game I'm destined to lose, until some arbitrary endpoint when I decide that it's time to move on and do something else. During those low moments, my faculties are impaired and my mind reaches a state of near paralysis. I fumble through conversations with patients, and I yawn repeatedly as medical students present cases on rounds. It's not uncommon for a cancer patient in unremitting pain to pause midsentence to ask me what's wrong. People who are dying routinely tell me I look exhausted.

It's genuinely mortifying to know you look like hell at work, yet I feel powerless to stop it. "It's happening," I'll say to Heather after we've put our kids to bed, and she'll know exactly what I'm referring to. On those days, I have a desperate desire to be left alone, removed from everything and everyone. I'll stare at the same page in a magazine, reading and rereading a paragraph again and again, waiting for the gears to move, thinking about a scared patient who is alone in a hospital bed, counting on me to help. I just have to wait for the fog to lift.

It's an odd position to be in. I've written about burnout in medicine, and I give lectures to young doctors who are worried that it might happen to them. I consult with residency training programs, providing tips for improving the culture of medicine while urging them to avoid hiring "wellness officers" and to skip the forced self-reflection. Physicians are stressed-out and beat-up, I tell them; what they really need is to feel normal again. They deserve sleep and a decent meal or an evening with friends. Maybe a glass of wine. They need things that a hospital cannot provide.

I like to watch documentaries. I was first interested in true crime specials—something I may have inherited from my parents, who both have doctoral degrees in criminal justice—and later I migrated to music documentaries. I had recently watched one about the Eagles with my close friends from college—a novelist, a screenwriter, a journalist, and an intellectual property lawyer—and we'd decided that I was most like the band's bassist and background vocalist, Randy Meisner. He was the gentle one, a forgettable figure in a band dominated by the outsized personalities of Don Henley and Glenn Frey, and I immediately felt a strange connection with him.

Meisner would lead a messy life—he battled addiction, heart disease, an imposter, a coma, and repeatedly threatened suicide—but he managed to cowrite and sing one beautiful, unforgettable song, "Take It to the Limit." The chorus had become a mantra of sorts, or perhaps the guiding principle I had been seeking. The song helped me push through the long days at the hospital and the longer nights writing about it. It made me smile every time I heard Meisner hit the high notes.

As the song played in my office, I scanned an article that I'd been meaning to read, locking in on a key statement that had been popping up in medical journals: antibiotic-resistant bacteria were becoming more aggressive. This was bad news for Donny and others like him, and helped explain why a small nick from a sheet of paper could cause an overwhelming infection. The rules of the game had changed, and for every success, it felt like there were two setbacks—or more. Before I had finished the depressing article, I heard a soft tapping. One of my colleagues, a new hospitalist, slowly inched my door open and cleared his throat. "I may have someone for you," he said. "Looks like a new case of cellulitis."

I paused the song and grabbed my white coat, my mood brightening immediately. A few minutes later, I was at the bedside of an elderly man named Louis. He was a retired cop, with a shock of white hair and a thin, wispy mustache, and he'd come to the hospital because his right leg had become red, swollen, and unbearably painful. Louis was eighty years old and still a hulk of a man, and he answered my initial questions with just a syllable or two. He wasn't the type to open up, or so I thought. "I see you're NYPD," I said as I reviewed his chart at the foot of his stretcher. "Thank you for your service." It was a vaguely awkward thing to say, but it was a habit of mine—something I'd picked up from a physician at Columbia Presbyterian who resembled the actor Scott Baio.

Louis put his hands behind his large head and smiled. "TPF," he said.

"I don't know what it says in my chart, but I was TPF. If, you know, if that matters somehow." I repeated the letters slowly to myself and scanned my brain for an acronym. "Tactical Patrol Force," he added. "You could say we invented stop-and-frisk."

"Hadn't heard of it," I said.

Louis explained that, decades earlier, he had been part of a small unit that was responsible for policing rough neighborhoods in the outer boroughs—places such as Brooklyn's Bedford-Stuyvesant and Fort Apache in the Bronx—by arresting the criminals, junkies, derelicts, and other undesirables. "We used to say, 'When the whistle blows, everybody goes.' If someone didn't look right, they were coming with us. Lock 'em all up. No questions asked."

"Huh."

He smiled. "We're the reason guys like you are moving to Fort Greene."

Louis had fond memories of his time with TPF, viewing his work as a genuine service to the community, but others, I learned, weren't so charitable. As he spoke, I typed those three letters into my phone. His Tactical Patrol Force had been described in *Vanity Fair* magazine as "an elite unit comprised of big, tough cops (there was a size requirement to join the squad), overwhelmingly white, who climbed fire escapes, traversed rooftops, and busted down doors with virtual impunity. In black and Hispanic communities of the early 1960s, they were despised, and among black New Yorkers of a certain age, they are remembered as a terror squad designed to instill fear in the community." It sounded a bit like rational drug design, but with human beings in the sauté pan. Communities were ripped open and torn apart in a misguided effort to make things better.

The man stretched out before me was a far cry from the freewheeling, hard-charging officer of the civil rights era. Louis could no longer even walk unassisted. The people of color who largely staffed our hospital could be the children or relatives of the folks he once encountered on the beat. He was the vulnerable one now, yet he spoke with the confidence of a man recalling heroic deeds, and with every anecdote, he became a bit less guarded. He lit up as he told me about the time he dressed up as a nurse, trying to get mugged, so he'd know whom to arrest. "Had the time of my life," he said. "We all did." Louis tapped his left hand on his chest, and I noticed that he was missing his ring finger. I imagined an altercation in an alley or a scuffle in a bar.

"Let me explain why I'm here," I finally said. "I'm running a clinical trial, and it looks like you might qualify." I pointed to his leg and described the risks and benefits of dalba before handing Louis the consent form. "I'll give you some time to read it over," I said. "I can come back later. I'm here all day."

"So am I, apparently." He scanned the paper and put it in his lap. "What's the point of this study? Why me?"

"The short answer is that we're studying a new drug. A new antibiotic."

"Tell me about it."

"It's for people with skin infections who require hospitalization. When other antibiotics fail, we try this new one." I pointed to the consent form. "The company is providing the drug for free so we can determine how best to use it. We hope it'll get patients out of the hospital safely and more quickly."

"But I don't want to be discharged. I want to be able to walk again." I put my hand on his knee, a few inches above the infection. I could feel the warmth radiating from his skin. My eyes fixated on the dark-blue blood vessels just beneath his fragile skin. His soft leg felt like tissue paper. "I need a tune-up," he said. "Not a quick discharge. I need to be able to walk. I need to see a physical therapist and a social worker. Keep me as long as that takes."

I nodded. "Got it."

He looked at the consent form and sighed. "I'm gonna pass on your trial," he said, gently pulling his leg away from me. "Thanks . . . but no, thanks."

"No problem."

I shook Louis's hand—the old man still had a firm grip—and assured him that he was in capable hands. As I was gathering my belongings,

Louis offered a final thought: his cardiologist had been encouraging him to write a book. "What do you think?" he asked. "Maybe I'll call it *When the Whistle Blows.*"

I took a moment to consider his question—perhaps too long—or just longer than he'd hoped. "Oh, no," he said, "don't tell me: you're a liberal."

I folded my arms. "Now, why would you say that?" I smiled to indicate that I hadn't been offended by the accusation. "What gave it away?" I looked down at my white coat and then up at his glimmering green eyes.

"Liberals hated us," he said. "We used to have the run of the town, and they hated it."

"Sounds like you've got quite a story to tell."

"My kids can't stand it when I talk about these things, but I got some good stuff. I was there when Castro came to the UN. Tackled a would-be assassin. Now, that—that was a friggin' mess."

I glanced down at my shoes and then asked him the question I want to ask all cops but am usually too shy to ask: "Ever shoot anyone?"

Louis grinned and pointed his right hand at me. Then he flicked up his thumb and extended his index finger toward my chest, like it was a tiny pistol. "No comment."

"Oh, come on."

"No chance, Dr. McCarthy. I'm saving that one for the book."

CHAPTER 28

Obstacles

I LEFT LOUIS feeling frustrated and a bit angry. I had been putting up flyers, emailing colleagues, and screaming from the treetops that a new antibiotic had finally arrived, one that might truly help people, and the company was giving it to us for free, but I had difficulty finding suitable patients who were willing to try it. Some of the men and women that I approached were *compos mentis*, but in subsequent days, many that I spoke with were sick and scared and unable to provide informed consent. Those that could were often excluded due to a technicality—something like a transient dip in blood pressure or an abnormal blood test—thus creating an agonizing paradox: the patients most in need of study were often ineligible.

I went back to my office and returned to the process of hunting for infected patients. I knew they were out there. My hospital sees thousands of skin and soft tissue infections every year. I just had to find them. After scrolling through pages and page of names, I had another hit. Jackson had returned to the emergency room with an expanding infection on his right forearm. He was in need of antibiotics and would be a natural fit for the trial. He was also a danger to others, harboring several superbug infections in his lungs and abdomen, and would have to be sequestered. Discharging him would be doing everyone a favor, and I could do it for free.

Jackson was in the room next to Louis, sitting with his wife, who was wearing a disposable yellow gown to protect her skin from infection. Before I entered the room, I retrieved his chart to make sure there was no evidence of bone infection, known as osteomyelitis, as this would be disqualifying. An X-ray had shown no evidence of osteo, but there was a different problem. Jackson's white blood cell count was dangerously low—the medical term is *neutropenia*—and he would need to be monitored closely in a special wing of the hospital for high-risk patients, known as the step-down unit. That floor had specially trained nurses and caregivers who could identify patients who were about to deteriorate and knew how to provide intensive care outside of an ICU. Jackson was sick—too sick—and was bound for intensive monitoring in the step-down unit, which meant he was ineligible for my trial.

Next to him, on a small wooden table, I could see a small bag of colistin. As our eyes met, a nurse entered the room, picked up the bag, and hung it on a thin metal pole just behind his stretcher. Years had passed since I'd first met Jackson—his dark hair had thinned, and he was slightly stooped now—and I wasn't sure he remembered me. I gave a small wave, and he motioned me into the room. There was an oxygen tank at his side, and he was wheelchair-bound. I soon learned that the man I had known as a mechanic had been unemployed for years, sapped of his identity by a cascade of worsening infections. He would probably never work again. I would need to think of him as something else, something other than a mechanic. "Good to see you again," I said, extending my gloved hand. "Although I wish it were under different circumstances."

Watching colistin drip into him wasn't easy; I could see that he was just as nervous as he'd been years earlier on that unseasonably warm day in October. We had made real progress since then—several new antibiotics had been approved, and more were on the way—but his case was a stark reminder of how far we still had to go. I'll never forget the terrified look on his face the first time we met; it wasn't much different the second time I saw him, or any of the times thereafter. "Good luck," I said after I explained why the study wasn't right for him. "You'll get through this."

He looked at the infected area on his arm and then up at me. "You think?"

Much has been done to streamline the process of enrollment, but as I was discovering, there were mounting obstacles, even for open-label studies like mine, where patients know what experimental drug they're receiving. At meetings around the world, my colleagues routinely complained that exclusion criteria had become too stringent, that conditions such as neutropenia and organ failure and sepsis—all of which Jackson was battling—were marginalizing patients in need of study. There were times when my eyes glazed over reading about the financial considerations of clinical research, but the takeaway was impossible to ignore: antibiotic studies were becoming too complicated and too expensive. I had heard whispers that Allergan was on the verge of reallocating resources to other areas of research—things such as Botox and eye disease—and I couldn't say I blamed the company. A study from the London School of Economics and Political Science estimated that, at discovery, the net present value of a new antibiotic was *minus* $50 million. My drug trial was secure, but others might not be.

I went back to my office and resumed the hunt for patients with cellulitis. After a few minutes, I heard a knock. I turned off the Eagles mix and stood up. A young doctor eased my door open and handed me a sheet of paper. "I think I have someone for you," she said. "Patient is allergic to vancomycin, and I think he has MRSA. Might be a candidate for dalba."

"Let me take a look," I quickly typed the patient's name into my computer.

"One question," she said as we stared at the man's name and medical record number. "Does dalba actually work?"

"I believe so, yes." It felt strange to equivocate, but, in truth, I didn't *really* know. "Yes," I added.

"It's just that this guy has been in the ER three times in the past month," she said, "and I don't want to give him false hope. I don't . . . I'm sure you get it."

"Let me go talk to him," I said, grabbing my stethoscope. "I can explain the risks of participating." My mind shot back to the months I spent drafting the initial protocol. The IRB had pressed me to describe in extreme detail just how dangerous the study could be—and with good reason. "I can also discuss the benefits."

I skipped toward the elevator, imagining how I would present the study to this new patient. As its doors were closing, a message arrived. Here was a bright spot: a reminder that it was time for the ContraFect site visit. The lysin protocol had sailed through the IRB, and the staph trial was ready to begin. Representatives from the company were coming to meet with our research team to discuss enrollment, informed consent, and administration of lysin to human volunteers. Vince Fischetti's vision might soon be realized.

CHAPTER 29

First

THE FIRST PERSON to receive dalba at NewYork-Presbyterian Hospital was a man who had become consumed by a single thought: *I'm drowning in quicksand.* Mark Simmons said those words to me from his hospital bed as we watched the morning sun creep above the East River on a balmy day in February. "For the past six months, it's just been a slow, gradual descent." He sucked briefly on his lower lip, searching for more words. "It feels like I'm turning into a pillar of salt." Simmons had spent his career as a litigator in Manhattan, protecting the interests of a large Fortune 500 company, and retired ten months before the symptoms began. "One day," he went on, "I'm riding my bike—"

"He rides it everywhere," his wife interjected. Janet was seated at the bedside, a mystery novel and newspaper in her lap, holding a tray of half-eaten pancakes. "You wouldn't believe how active he is. Was."

Mr. Simmons used his arms to swing his legs off the side of the bed and sat upright, jutting out his chin. He was a thin man, just over sixty years old, with hazel eyes and thick, white hair. "One day I'm pedaling downtown," he said, "and all of a sudden my legs start to feel heavy."

"A week later," Janet said, "he's wetting the bed. It all happened so fast."

"I've been in diapers for months."

Simmons saw his primary care doctor, who didn't know what to make of it, and told him it was simply a part of getting older. "I wasn't buying it," he said as he reached for the tray of food. "So I did some searching and found a new doctor."

That physician, a movement specialist, was concerned that Mark was showing early signs of a progressive neurologic condition such as Parkinson's disease or normal pressure hydrocephalus, known as NPH. In medical school, students learn that people with NPH exhibit the three W's: they're wet, wobbly, and weird. Patients wet themselves, have trouble walking, and may exhibit bizarre behavior. The triad is a nice memory device, one that has stuck with me for nearly two decades, but the amusement dissipates once you meet someone who's been stripped of his dignity by the disease.

Mark was put through a battery of tests and eventually given a trial of carbidopa-levodopa, a variant of the chemical dopamine, to treat his worsening symptoms. His doctor thought he might have Parkinson's disease, which occurs when nerve cells don't produce enough dopamine, leading to muscle rigidity, tremors, difficulty speaking, and, in some cases, an expressionless face. The treatment didn't work. He continued to feel trapped, a prisoner in his own skin, wading deeper and deeper into quicksand. "Sometimes I wake up screaming," he said. "Thinking I'm about to suffocate."

His wife nodded. "It's horrible."

"I kept thinking, *My God, I'm ossifying*." He glanced at Janet and repeated the word again. *Ossifying*. After I asked to see his infection, Mark rolled onto his side, pulled up his hospital gown, and exposed his bare buttocks. He had a scarlet rash just below his tailbone, covering the entirety of his right gluteus. When I merely touched it, he flinched as if he'd been spanked.

I noted a small tear—an area where the inflamed skin had split apart—and wondered if his urine was seeping into it at night. "The symptoms keep moving up," he said. "First the feet, then the legs, now the thighs. I can't move, I can't get up to pee, I can't do anything." I tried to gauge how the infection might respond to dalba. The truth was that I wasn't sure.

"Do you think he'll ever ride a bike again?" Janet asked.

This was a question I could not answer. "I don't know," I said. I locked eyes with her, as I always did with patients or their families after acknowledging uncertainty. "But this is a great place to get an answer."

Janet brightened ever so slightly. "We have an appointment with a neurologist here next week," she said. "Fingers crossed." They were understandably focused on the big picture, but I was down in the weeds, thinking about his skin, and the urine that might seep into it and make the infection worse. Hospitalization could be a dangerous misadventure for Mark: he might end up with a urinary catheter, and, with it, the attendant risks of infection.

"I'm here to talk about a clinical trial," I said as I reached for my business card. "It's an antibiotic study. A drug trial." If any patient was going to understand the nuances of informed consent, it was Mark Simmons. I'd never met a lawyer who would blindly sign a document, even in distress.

"This drug any good?" he asked. He put on his bifocals and started reading.

"Well," I said, "you'll be the first to receive it here. The first in this hospital."

Janet frowned. "You've never given it to anyone?"

"I have not."

She folded her arms. "So why Mark?"

I explained the risks—the antibiotic was new to me and my colleagues—and the potential benefits, just as I had practiced. I had spent six months observing all of the things that could go wrong during a hospitalization for a skin infection and was eager to serve as a catalyst for change. With dalba, we could avoid unnecessary tests, possible blood clots, and exposure to all kinds of dangerous bacteria. "Why don't I give you some time to think it over?" I said. "I'll come back."

"I really just have one question," Mark said. "Would you give this drug to your own mother?" He and Janet looked at me. I had spent years thinking about dalba, considering all sorts of things about the drug, but I hadn't asked myself this. It was a terrific question. "I would," I said. "Yes, I would give it to her."

Mark put down the paper and clapped his hands. "Okay," he said. "Let's do it. I'm ready to get outta here."

My heartbeat quickened when dalba entered the room an hour later. The drug was hand delivered in a small, clear bag from the investigational pharmacy. The porter carried it with two hands, palms up, as if he were holding something precious—which he was. It was the first time a patient at NewYork-Presbyterian Hospital would receive the drug. I held my breath when the clear solution started dripping into Mark's arm, and tried not to think about Omniflox. The dalba infusion lasted thirty minutes, during which time I monitored his vital signs and made small talk with his wife. I did not mention that I was required to be there in case something catastrophic happened.

I tried to make it seem like this was all routine, and that I wasn't monitoring every muscle in his face, or that I didn't cringe every time he took a deep breath. Mark probably didn't know that I was keeping track of how frequently he blinked or that I had muted the television to listen for subtle signs that his throat might be closing. His nostrils flared twice during the infusion, and both times I took a quick step in his direction, only to withdraw when his nose returned to its normal shape. When the infusion was over, I shook Mark's hand. He still had a sturdy frame and a strong upper body. "That's it?" he asked.

"That's it."

"I can go home?"

"Yes. As soon as your doctor gives the okay, you can go."

Treating his infection would be a small victory. Mark still felt like he was wading in quicksand, and it was only getting deeper. I had done nothing to reverse those symptoms, but perhaps someone else could. Maybe someone else could get him back on that bike.

"Now let's talk about follow-up," I said. "What's the best number to reach you?"

CHAPTER 30

Alicia

I LEFT MARK'S ROOM feeling relieved—I had finally administered dalba—but the moment was fleeting. There was no time to fixate on the unknowns surrounding his case. I had to pivot from the clinical trial to my patients on the general medicine floor. I was now tasked with evaluating a twenty-six-year-old woman named Alicia, who had come in complaining of crippling body aches that had been worsening over the past year. I had been told that she had arrived with her frustrated father and a binder full of test results. "Good luck," the triage message said.

I made a note to call Mr. Simmons and took the elevator to the fifth floor, where I started the process of reviewing her records. The documentation was brief but concerning: Alicia had been to six hospitals in eight months trying to find out why it felt like her skin was on fire. She also wanted to know why stress caused her to pass out. Diagnoses such as chronic regional pain syndrome, lupus, muscle inflammation (myositis), and a personality disorder had been considered, but none of them had stuck. Alicia had flown to San Francisco for yet another opinion, but doctors there were stumped, too. So she came to our hospital searching for an answer, and it was my job to give it to her.

Alicia was resting under a pink blanket when I entered her dark room. She looked a good bit older than her age. She was gaunt, with long, blonde hair and large, brown eyes, and there was a solitary crease running across her forehead, giving her the appearance of someone deep in thought. The lights were off, but the television was on, and her dad, wearing a Buffalo Bills sweatshirt and a Yankees hat, was seated in a plastic chair next to her bed.

Alicia threw off the blanket and pulled out a sheet of paper. "I have three questions," she said. She took a deep breath and exhaled slowly. "One, do you think I have Lyme disease? Two, do I have a chronic mold infection, and three, do I need a PET scan?" She held a pen close to the paper, prepared to record my answers. "You can answer in whatever order you'd like."

Alicia's father looked at me expectantly. "Anything you can do," he said softly. "We would be grateful." As I considered her questions and how I might answer them, my mind traveled to a story I had heard a few days earlier. When Tom Walsh's daughter was a high school freshman, he was asked to write a paragraph about her for her school. It was intended to be a brief statement that captured something essential about his child, but when he was finished, Tom realized he'd written nine exuberant pages. Looking at Alicia and her anxious father, I thought of that letter and the unique bond between a father and daughter. It had been on my mind a lot lately—media coverage of school shootings had something to do with it—and when I tried to imagine caring for a sick child, it was almost too painful to think about. "I'm going to do my best," I said.

Alicia's father reached under his chair to retrieve a purple binder and handed it to me. "It's all in there," he said. "Every test." He had a worried look on his face, and soon I did, too. There were hundreds of pages documenting millions of dollars in tests and consultations with specialists from around the country, yet here they were without answers. "I know it's . . . a lot."

"May I take this?" I asked, holding up the binder.

"I'd prefer it if you didn't," Alicia said. "I don't want to lose it."

"Understood," I said, flipping through the tattered pages. "We can photocopy it, if that's okay."

"Sure."

"So," I said, pulling a chair to within a few inches of her bed, "how did things go at the last hospital?"

"Badly," she said. Alicia cracked her knuckles and looked at her dad. "My fingers are killing me."

"And the one before that?"

"Same. Doctors eventually just give up." Her eyes were damp. "That's what happens every time. But I'm still here. Still hurting."

I leaned forward in my chair and continued to flip through the pages, noting that many tests had been repeated several times. Every hospital seemed to start from scratch, initiating a million-dollar workup time and again. "You've been tested for Lyme and mold," I said. "And both were negative."

"Yes, but . . ." She shook her head. "Are you familiar with chronic Lyme?"

It was a controversial condition, one that led many patients to antibiotic therapy they might not need. I could see that Alicia had received a half dozen drugs for conditions she appeared not to have. "You were treated for Lyme?"

"Many times," she said. "Test was always negative, but I think I had it. I'm sure of it."

"And what happens when you're treated? Do the antibiotics help?"

"Honestly," she said, again looking at her father, "it makes it worse."

"And mold? You were treated for that, too?"

"Yes."

Her father cleared his throat. "So was I. It didn't help." He held up a little bottle of pills and shook it. I had cared for large cohorts of patients with mold infections and even more with Lyme disease, but none of them was like Alicia. Those patients tested positive for the diseases and weren't nearly so debilitated. I closed the binder and thought about whether I should bring up an alternative diagnosis, one that might provide some clarity. "I noticed that a psychiatrist saw you in California. Someone who mentioned untreated depression. Is that something you want to talk about?"

In a flash, the tenor of the room changed. Alicia's lip started to tremble, and her father stood up. He took the binder from me, turned off the television, and walked out of the room. "It really upsets me that you just said that," she said. "I'm coming here for help, and you haven't even examined me. You just . . ." She closed her eyes, unable to finish the thought.

It occurred to me, in that difficult moment, that the binder was her father's nine-page letter. It represented the many months they'd spent together traveling from one hospital to the next, exchanging small talk, waiting for tests, searching for doctors, aching for some explanation. It was his expression of love for his daughter. And with one word—*depression*—I had extinguished whatever bit of hope they may have had. "Yours is clearly a difficult case," I said, fumbling for words, "and I have to consider all kinds of diagnoses, even unlikely ones. Everything's on the table."

Alicia shook her head. "I would like you to leave, Dr. McCarthy. Now. Please, go."

CHAPTER 31

Persuasion

THE NEXT PATIENT to enroll in the dalba trial was a twenty-nine-year-old security guard who worked at a nearby hospital. "I came here," Gerard Jenkins said, shortly after Alicia had excused me from her room, "because I didn't want anyone at work to see this." He pulled up his pant leg and showed me a large, red rash, one that looked like so many of the others I'd seen during my year chasing infected patients. An ER physician had taken a marker to the area to record the boundaries of the rash, which I would use to monitor his progress. With proper treatment, the infection would gradually recede from the blue outline. After I explained the purpose of my study, Gerard posed a single question: "Can you give me a doctor's note? Something that says I need to stay home for a while. Like, an extended period."

Gerard understood how things worked. As long as he was hospitalized, his employer knew that he couldn't perform his duties as a security guard. But once he was home, the clock started ticking. Dalba might hasten his discharge, but it might also hasten his return to a stressful job. Hospital security is difficult work—patients can be aggressive, belligerent, and occasionally violent—and it requires a degree of fitness and agility. It's not the kind of thing you can do on a bum leg. Gerard pointed at the consent form and then at his infection. "I need to be out for a week, don't you think? Or more?" He flashed a gap-toothed grin. "Probably more."

Gerard was proposing a trade: he would volunteer for the trial if I would help him get a few extra days off work. It was wrong, clearly, but that didn't stop me from thinking about it. He was unwell and needed time to recuperate, but how much time was debatable: some people need a few days, others need far longer. Regardless, it was inappropriate for me to participate in that debate. "One of my kids," he added, "he needs me at home. I'm sure you get it."

I sighed. "Unfortunately, I can't do it," I said. "I can't write the note. But you can ask your doctor about it. I bet you'll be able to work something out." The bright line between medical practice and research is often blurred, and this gentle form of coercion is not uncommon in clinical trials. As financial pressures increase, investigators occasionally feel the need to push boundaries to stay within budget, exchanging favors for informed consent. Writing that letter wouldn't cost me anything, but it would compromise the integrity of the study. "I just can't." My eyes focused on the space between his two front teeth and then moved to the small cleft in his chin. "I'm sorry."

Gerard winked at me. "I get it," he said. "Figured it was worth a shot." He looked at the leg and then back at me. His eyes were heavy—he had been in the emergency room for seventeen hours—and he'd had the misfortune to share a room with a demented man who was up all night screaming about lightbulbs. "So," Gerard said, "where do I sign?" He reached for the bedside remote and started flipping through the channels. "I'll do the trial. Bring it on."

Gerard was a vulnerable patient—a guy making just over minimum wage, with three kids—who easily could be exploited, and I felt uneasy. He was exhausted, and I wasn't entirely sure he understood the details of the study or what I expected to accomplish by treating him. I thought about the town and the trial that were forever imprinted on my consciousness. "Let me make sure you know the full story," I said. "With this trial."

We had come so far since Tuskegee, implementing layers of oversight and safeguards that prevented the most egregious ethical violations, but we still faced challenges. There were still men and women who could be manipulated and ambitious trials that might exploit them. But to exclude these patients would also do a disservice. A study from the nonprofit investigative journalism website ProPublica found that black patients are underrepresented in clinical trials of new drugs, even when the treatment is aimed at a disease that disproportionally affects them. "There's a lot to go over," I added, "but it won't take long." Marginalized patients need to be represented in clinical trials, just like everyone else. To exclude Gerard based on demographics would remove him from scientific progress. To generate knowledge, we need data, and, for better or worse, that must come from human experimentation.

Ninety-four-year-old Herman Shaw, victim and survivor of the Tuskegee Syphilis Study, speaks as US President Bill Clinton looks on during ceremonies at the White House

Gerard listened as I flipped through the pages, explaining how and why the study was being done. He nodded when I nodded, and wrinkled his brow when I did. "Sounds good," he said when I was finished. "Let's do it."

"You would sign here," I said, handing him the papers.

An hour later, Gerard and I watched silently as dalba dripped into his body. After the first few drops, he chuckled. "I'm like a guinea pig," he said. "Just a big ol' pig." I imagined the drug traveling through his bloodstream and down to his leg, encountering millions of bacteria just under his skin. Dalba would prevent those bugs from building cell walls, which would keep the infection from spreading. His immune system would take care of the rest. At least, that was the plan. "This is kinda exciting," he added.

I patted his shoulder and gave it a squeeze. After a few minutes, Gerard closed his eyes. Soon he was asleep. As he snored, I walked toward a large window overlooking Roosevelt Island and thought about my other patients. I had dozens of calls to make and clinical notes to write, and I would need several hours to review all of Alicia's records. I'd probably listen to "Take It to the Limit" fifty times just to get through it all.

It had been a mistake to mention a psychiatric condition to Alicia so quickly and to do it after closing her binder. It must have seemed like I was discounting those tests—the ones that meant so much to her—and suggesting she was crazy. I would have to go back and apologize. My reaction had been cruel, like tearing up Tom's nine-page letter in front of him.

The exchange exposed a hollow conceit: I was Alicia's ally—a doctor firmly in her corner—up to a point. I was willing to sift through hundreds of pages of documents and consult countless experts to evaluate and adjudicate her case, but if she wasn't willing to follow my advice or speak to those consultants, what was the point? I needed her to buy into my way of thinking; otherwise, we had no alliance. I would apologize, but I would also ask her to believe in me.

I was typing a text message to a medical student when an alarm sounded next to Gerard's bed, indicating that the infusion was complete. He opened his eyes, packed up his belongings, and called his wife. "Coming home early," he said into the phone. "No, no," he said, looking at me. "Today." A half hour later, after Gerard's doctor had given him a discharge summary, I walked him to the elevators and pointed him in the direction of the lobby. "See you in a couple of weeks," he said. "And don't forget to call me!"

I gave him a thumbs-up. "I won't. I definitely won't. See you then." When I returned to the hospital ward, I saw that a cleaning crew had already entered Gerard's room to prepare it for the next patient.

CHAPTER 32

The Rollout

A FEW DAYS LATER, I received an email from an assistant professor of medicine who cares for patients on the general medicine floor, with the subject line "The word is getting out." I bit my lower lip as I opened it. My colleague had been reviewing the chart of a twenty-six-year-old woman who had arrived in our emergency room in the middle of the night with left ankle pain, redness, and swelling that hadn't improved with the antibiotics cefadroxil and Bactrim. "When I saw the patient," the doctor wrote to me, "she said she did not want to stay in the hospital. She wanted to leave."

The doctor had heard about my trial and mentioned dalba to the patient. "I felt that she would be a good candidate for that drug," he wrote. I smiled when I read those words. It felt like a turning point, one that had taken years to reach. I had given dalba to only two patients, yet doctors I barely knew had become aware of it. They understood who might qualify, how the drug was metabolized, and where to find it. For nearly a year, I had been chasing patients; now their doctors would be coming to me.

I couldn't wait to tell Tom. After morning rounds, during which I attempted to repair my fractured relationship with Alicia and her father, I glided over to his office to show him the email, but he wasn't there. I pulled out my phone and stared at my calendar. It had been synched with his, but I still routinely lost track of him. *Thursday.* I knew where I could find him.

Tom and I spend every Thursday morning in the clinical microbiology laboratory reviewing the hospital's most interesting cases of the week. It's a chance for us to peer into microscopes and examine superbugs head-on, just like they did in the old days, and to speak with the pathologists and technologists who do this full-time—the ones who know when danger is coming before everyone else. The men and women in the microbiology lab notice subtle shifts in antibiotic resistance patterns and changes in the way pathogens spread.

"We're not the backroom boys anymore," one microbiologist in my department likes to say to me. And he's right: the soft-spoken professionals who stare into microscopes for a living have been thrust onto center stage. They tell us when a new efflux pump has emerged and when a new porin mutation has arrived. They tell us more about superbugs than just about anyone else. "These people are our friends," Tom says of the microbiologists. "If you don't know their names, learn them."

Once we've exhausted our eyeballs on Thursday mornings, we head back to Tom's office to review what we've seen and consider whether any of it merits further study. Ideas volley back and forth, and clinical trials are hastily sketched out on notepads. Hundreds of ideas are proposed and rejected, both usually by him, before we settle on a project and begin the hunt for funding. Emails are dashed off, phone calls are made, and textbooks are pulled off the shelves. The meetings are invariably interrupted by unexpected visitors, musical interludes, and random asides that have nothing to do with antibiotics. Nothing to do with anything, really.

Sure enough, on this particular Thursday morning, I found Tom in the laboratory, chatting with an Eastern European technologist. Tom shook my hand and asked me to look into the microscope. "What do you see?" he asked. "Top right corner . . . Do you see it?" He was exquisitely observant—the type to spot a four-leaf clover in a field—and he was drawn to microscopic phenomena. "Look closely."

My eyes eventually settled on something that looked like a tiny banana or possibly a canoe. "Macroconidia?" I asked. It was a word I had never heard during medical school or residency training, but I found myself using it all the time. It was a small piece of a fungus, one that helped it reproduce. I looked up from the microscope. "Looks like a mold," I said. "Not a black mold, something else."

Tom nodded. "What you're looking at," he said, "is from the finger of a man with AML." Acute myelogenous leukemia. "Any guesses?"

"*Fusarium*," I said confidently. "Has to be." The first case of it that I'd ever encountered had been with Tom seven years earlier. I wondered if he remembered. "*Fusarium solani*." The fungus infected patients with cancer, traveling through the bloodstream to the skin, forming deposits on relatively cold surfaces such as fingers and toes. Doctors often mistake it for bacteria.

"Correct, Professor!"

I pulled out my phone and took a picture for my records.

"This fungus is becoming quite difficult to treat," Tom said, noting that two antifungal agents were often necessary. Our hospital had to ship the organism to another state just to find out which drugs might work. *Fusarium* wasn't listed as a superbug yet, but it might be soon. "The patient said he developed the infection shortly after a paper cut."

My eyebrows bounced with those last two words. It had to be Donny, the first responder who'd hastily grabbed his gear after the Twin Towers fell. His skin infection hadn't been from bacteria after all. It was from a fungus. "I know this guy," I said. "Doing okay?"

"He's doing much better now," Tom said. "Finally got him on the proper treatment."

While I was thinking about Donny, an infectious diseases specialist walked in carrying a blue-green petri dish. "Interesting case," she said. "Young healthy guy comes in with back pain, turns out he has an aortic aneurysm." The man had developed an abnormal bulge in the largest and most important artery in the body, the aorta, leaving the blood vessel weakened and prone to rupture—an event that could kill him.

"This," she said, pointing the plate toward us, "is from the biopsy."

We looked under the microscope at a collection of bacteria while the doctor gave us a few more details about the case. "The patient travels a lot for work, mostly to the Southwest. Says he eats a lot of local food."

Tom leaned back from the microscope. "*Salmonella*," he said. "May I see the plate?" He noted the lawn of black bacteria covering the petri dish. "This is classic," he commented, launching into a discussion of organisms that infect large blood vessels such as the aorta. A recent gene mutation in *Salmonella*, known as a single nucleotide polymorphism, or SNP, had created a species of superbug that was killing hundreds of thousands of Africans every year. A similar outbreak of extensively drug-resistant (XDR) *Salmonella* was spreading through Pakistan. It was only a matter of time before it moved to other continents.

Salmonella was associated with food poisoning and grouped with other neglected tropical diseases that the Bill & Melinda Gates Foundation might want to address, but it didn't normally raise alarm bells for the rest of us. That was about to change. "This can be lethal," Tom added. "Never forget that." We had seen a devastating case in a patient who had received a stem cell transplant; the infection was transmitted from a pet lizard that had been allowed to lick the owner's ear.

"How's this patient doing?" I asked, staring at the plate of *Salmonella*. "Going to the operating room?"

"Just had surgery," the woman said. "They did a femoral vein graft. We think he's going to pull through."

"Have you identified the source?" Tom asked.

She held the petri dish up to the light, and we gathered under it. "There's an outbreak in Colorado right now," she said. "Probably got it from a burrito. Nearly killed him." We all shook our heads in disbelief.

Tom and I had just returned from the microscopes to his office when I finally showed him the dalba email. "Word is getting out," I said, repeating the subject line. I could hardly contain my excitement. While we were discussing *Fusarium* and *Salmonella*, I had received two more text messages from doctors seeking to enroll their patients. "People are lining up for it," I said. I had been in close contact with the patients who enrolled in the trial, including Mark and Gerard, ensuring that they had recovered safely at home, and made sure their doctors knew how dalba worked. This caused a ripple effect as physicians and patients in the community discussed the new treatment option. In a matter of days, patients were coming to my hospital requesting the drug, and clinicians were asking me about off-label uses.

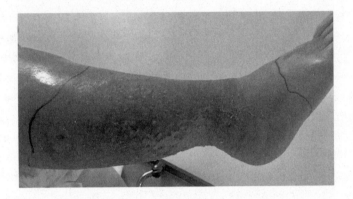

The right leg of a patient with a skin infection

Tom smiled and put his hand to his chin, doing a perfect if unintentional imitation of that iconic photograph of Steve Jobs. "Now *we* need to get the word out," he said. "On a larger scale." I took off my white coat and pulled out a pen. "We need to present the data," he went on.

We agreed to submit our study to one of the largest gatherings of infectious diseases specialists and researchers in the world. These conferences are always a hectic few days; Tom and I catch up briefly with old friends but spend much of our time with representatives from Big Pharma and biotech firms, trying to secure funding for the next project. I also sit in on seminars, listening as scientists and clinicians describe their latest findings, but this time things would be different. I would be doing the talking—getting out the word that Tom and I had developed a model of clinical research that really worked. With minimal funding, he and I could design studies that impact hospital length of stay, or some other financial metric, and I could evaluate the value of their compound in real time. The IRB process had been painful, but I had learned from it. I knew I could design a protocol, usher it through regulatory hurdles, consent patients, administer the drug, and arrange follow-up. I could show that an expensive new antibiotic could pay for itself. It almost sounded easy.

Many academics were already doing this type of work, of course, through rigorous clinical trials or mathematical modeling, creating projections for how and when a drug should be used. But those analyses don't always replicate reality. Much like the models for global warming, there's always disagreement about how best to interpret forecasts, and some hospital formulary committees—the experts who evaluate the safety and effectiveness of a new drug and determine whether it should be used—want to see real data, not models. They want to know that a new drug is not the next Omniflox.

"I suppose I shouldn't be surprised," I said, still basking in the glow of our early results, "but I figured we were gonna hit some roadblock. Something to derail the whole thing."

"Why would you think that?" Tom asked. The truth is, I was always expecting disaster to strike, and had to stifle that sentiment at the bedsides of frightened and anxious patients. I tried to exude confidence and competence, even when those things were lacking. But with Tom I could be myself.

"I don't know," I said, shrugging my shoulders. "I was expecting some snag during the rollout." I thought back to the months we'd spent haggling with the IRB. "Guess I'm pleasantly surprised." I scribbled my name in the margin of my notebook, and under it wrote "the little engine that could." Then I made a note to ask him about Alicia's case.

He put down his marker. "Are you familiar with Sun Tzu?" he asked. Tom was referring to the author of an ancient Chinese military text on warfare and leadership; I had once heard the New England Patriots head football coach, Bill Belichick, reference it.

"Vaguely."

"He said something that I think about quite a bit," Tom said, "especially when I'm running a trial: *The battle is won before it is fought.*"

CHAPTER 33

Investments

ALL EYES WERE ON TOM. We were seated around a large conference table next to his laboratory, shortly after I'd given dalba to my third patient, and the well-lit room had fallen silent. We were there for the weekly transplant meeting, where faculty members gather to discuss their most difficult cases, and when they're stumped, they turn to Tom. He was sitting at the head of the table with his arms folded. "So I'm not sure what to do," a physician said from the other end of the table. "And time is running out."

The case in question was a young mother who had just received a stem cell transplant and was now clinging to life in the ICU. She had developed an infection in her blood with a bacterium called *Enterococcus*, and it had just invaded her heart valves, causing a dangerous drop in blood pressure. This organism happened to be resistant to vancomycin, the antibiotic discovered by the missionary in Borneo, which gave the bug the name vancomycin-resistant *Enterococcus*, or VRE. They were three letters that we had all learned to fear.

VRE was first detected in England and France in 1986. A year later, it was found in the United States, traveling all the way from the Siouxland region of North Dakota to downtown Detroit. I saw it sparingly as a medical student, but now I encountered it every other week. Our hospital was always on the lookout, screening high-risk patients with a rectal swab, searching for those who might carry VRE on their skin or in their intestines, unaware that they were harboring a deadly pathogen that could easily spread to others. "If anyone has any suggestions," the doctor said, "I'm all ears."

Treatment options were shrinking for VRE, and they're especially limited for patients with weakened immune systems and little margin for error. How it had reached the stem cell transplant recipient at our hospital was unclear, but from the tone of the presentation, it was obvious that death was on the horizon. The medical team needed Tom's help. "Daptomycin?" the doctor proposed. A versatile antibiotic—discovered in a soil sample from Mount Ararat in Turkey—it was one of the few available options for VRE, but he wasn't sure it would be effective. "High-dose daptomycin?" he asked, looking around the room.

Tom shook his head. "That won't work," he said. "The organism is resistant." It was once an unthinkable scenario, but the woman's VRE infection had mutated to evade daptomycin, too, transforming into a superbug with almost no treatment options. I had heard of a specialist at another hospital treating it with Synercid, a combination of two antibiotics—quinupristin and dalfopristin—that become more powerful when used together, but it was an off-label use, and there was no guarantee it would work. Besides, our hospital didn't carry it. The room waited for Tom's pronouncement.

These were the most stimulating moments in medicine, the flash points when experts were stumped, and I had a front row seat. There was the case of *Scopulariopsis* in Denver, the *Saprochaete clavata* in Germany, and now this: a case of daptomycin-resistant VRE endocarditis in a transplant recipient with three small children and a husband who worked nights. The doctor at the head of the table was baffled, just like the rest of us. I wondered if Vincent Fischetti had some spare lysin in his laboratory at Rockefeller, something that would make *Enterococcus* explode.

"Let's back up," Tom said. "What's the differential on the white count?"

I checked my phone—I was due to see Gerard Jenkins later that day for his two-week follow-up—and typed in a reminder to ask him if he ever got that doctor's note excusing him temporarily from work. We had spoken twice since his discharge. The infection on his leg had disappeared after just three days. He wanted to know if I could examine him at work or if he would need to take time off for our follow-up. I sent him a quick text: "Best for you to come here."

Tom cleared his throat. "This patient does not have endocarditis," he declared. He surveyed the room, making eye contact with each transplant doctor, and concocted a new treatment regimen. "You should always question a diagnosis of endocarditis in a patient with neutropenia." We all scribbled down the dictum—it was the first time I'd heard it—and then moved on to another case. (He was right, of course: researchers had put forth a number of theories to explain the endocarditis phenomenon, but the science wasn't settled yet.) "What's next?"

When the meeting was over, Tom pulled me aside. "Do you know what you're going to say?" he asked. His phone had buzzed throughout the conference, but he hadn't checked it once. I imagined him scrolling through the hundreds of unread messages late into the night, trying to parse the urgent from the semi-urgent. Digging into those correspondences was like going deep underwater, he said, and the process of responding was like slowly coming up for air. Once a critical mass of them had been addressed, Tom would say that he'd reached periscope depth.

"I do," I said. We were scheduled to sit down with a team of administrators from our hospital's finance committee to discuss their investing in a machine called T2Candida, for diagnosing fungal infections. Tom and I had been working with T2Candida for just over a year, collecting blood samples from patients in the ICU, and we discovered that it could detect pathogens in blood faster than standard hospital methods. It was a complicated machine, relying on the same technology used by MRI scanners to find microbes in the blood of patients with unexplained fevers. It was also expensive, and it was our job to convince the suits that the investment was justified. "I think they'll see the light," I said.

Much of our superbug work focuses on antibiotic development and clinical trials, but diagnostics play an equally important role. Better tests mean more accurate diagnoses, and that, in turn, translates into

better use of antibiotics. Patients are constantly exposed to drugs un-
necessarily, and it usually happens when there is diagnostic uncertainty.
We try to remove the question marks, giving doctors the confidence to
stop antibiotics when they're not needed, and we do it with nifty gad-
gets. But someone has to pay for it.

Tom was a part of the old guard: he cared for patients until they were
better, regardless of how long it took or how much money was spent. The
mission came with no strings attached. I told him I was going to open
the T2Candida pitch with three words: *length of stay.* "Patients are waiting for days in the ER," I said softly, practicing my pitch as the transplant
doctors filed out of the conference room. "We can help."

A moment later, a new team of administrators came in. Tom stood
up and started the introductions. I was seated next to Ryan, a financier
who said he was on the hospital's quality improvement team, and next
to him were representatives from the microbiology lab, who were performing much of the legwork for the study. We all shook hands.

Ryan welcomed several more people who were joining via conference
call as we exchanged business cards. "Let's get started," he said. Tom
passed around a one-page summary I had written that explained the
rationale behind our T2Candida project. I had done my best to summarize the technology in layman's terms, mentioning that the machine
could detect subtle changes in the configuration of water molecules,
which allowed it to identify pathogens in blood.

I spoke up. "Let's talk for a moment about something that's on everyone's mind: length of stay." It was something we might not have opened
with a decade earlier. But times had changed, and the popularity of
dalba was proving how valuable that phrase could be. I quickly ceded
the floor to Tom so that he could explain the technology and why it was
a tremendous improvement over other diagnostic methods.

"Where does this fit into the guidelines?" someone in a navy pantsuit
asked. "Is this a new standard of care?" She appeared to be unconvinced.

"People don't come here for guidelines," Tom said. "They come here because it's one of the best hospitals in the world. They come here for cutting-
edge medicine. And that's what we're providing. That's what we do best."

"The FDA has approved T2Candida," I said, "and we anticipate FDA approval of T2Bacteria in the coming weeks." It was a guess, but I had heard that the FDA was about to render its decision, and the verdict would be favorable. "We talk about patient-centered care. Well, this is it."

"No one else in our network is using this," Ryan said. "Not that it's a problem; just something for us to be aware of."

"The implications are far more broad than length of stay," I said. "Faster and more accurate diagnosis means better use of antibiotics, and that saves lives, prevents the emergence of superbugs and saves the hospital money." The muffled sounds of two voices talking came through the speakerphone, but their words were unintelligible. "The machine we have now is on loan," I added. "Soon we'll have to give it back." I pointed at the one-page summary. "We'd like another."

Tom picked up the paper. "The data are clear," he said. "This technology will save lives."

"There's a young mother with VRE in her blood," I said, recalling the case we had just discussed. "Someone in this hospital right now who might die from the infection." I looked slowly around the room, settling on the skeptical woman in the pantsuit. "We're trying to identify these patients before it's too late."

Ryan jotted down a note and passed it to the person seated across from him. Then he typed something into his phone and looked up at me. "I love it," he said. "This is good for us and good for patients. Let's make it work."

CHAPTER 34

Into the Haystack

THE TROUBLING CASE of the stem cell transplant recipient with the VRE infection highlighted yet another problem in our fight against superbugs: the discovery of antibiotics is wildly inefficient. Many of our best drugs, including vancomycin, daptomycin, and nystatin, were discovered in soil samples by scientists who had the good fortune to look in the right place. But relying on serendipity is risky business. Despite nearly a century of experience with antibiotic discovery, we still don't really know where to look or how best to isolate the molecules beneath our feet. Is Borneo better than Bora Bora? What about the desert? Or the ocean? We hunt for antibiotics in sewage and polluted lakes and the intestines of insects, but the results have been inconsistent. There's a lot of dirt to sift through. We need a better way, or at least a more focused process, to identify the next big thing.

The day that I learned about the woman with the VRE infection, I started searching for better answers. It was difficult to accept that there were so few options at our disposal or that a team of transplant doctors had to turn to Tom to find a path forward. Big Pharma had invested billions of dollars to combat infectious diseases, yet this young mother might die because no one had found a drug to help her.

A few hours into my hunt for treatments, I stumbled upon a study by a scientist who might actually be onto something. Sean Brady is a micro-biologist at Rockefeller University with a PhD in organic chemistry and a laboratory near Vincent Fischetti's group. His team showed that dirt from Prospect Park in Brooklyn harbors genes that can produce more than two dozen new drugs. Using a method that couples DNA sequencing and bioinformatics, his group examined DNA extracted from more than two thousand soil samples taken from across the United States and found something exciting: a new class of antibiotics. Brady dubbed his discovery "malacidins," short for metagenomic acidic lipopeptide antibiotic-cidins, and he showed that they were able to kill all kinds of bacteria, including MRSA.

His crucial insight was to narrow the search: rather than looking for one compound at a time, he used a computer program to hunt for the DNA signature of calcium dependence across samples from all kinds of climates. It was an elegant approach, one that seems so simple in retro-spect, as the best ideas often are. As I read Brady's paper, I scribbled a note to myself to learn more about his process and underlined a key finding: after three weeks of exposure to malacidins, MRSA showed no signs of developing resistance. Brady's new drug had stumped the bug.

The work confirmed what we've always known to be true—dirt is a great place to find antibiotics—and his team figured out how to find the needle without sifting through the entire haystack. When they tested malacidins on MRSA skin infections in rats, the rodents experienced no side effects, suggesting the compound might be safe for testing in hu-mans. I wondered if there was an opportunity for collaboration. Malaci-dins prevent bacteria from forming cell walls, but humans use a different process, one that isn't affected by his antibiotic. It appeared safe for pa-tients, at least in theory. Perhaps it could help the mother with VRE.

These are the moments that put a kick in my step, when a project is first coming together and the opportunities seem limitless. On those days, my pessimism recedes, and I am lost in unending thought, dazed by ideas. During dinner, my kids will tap me and ask, "What are you thinking about, Dad?" Sometimes I'm not entirely sure, but this time I could see the trial taking shape. Brady had done the hard part, finding the drug, but Tom and I had access to patients—hundreds if not thousands who were in need of treatment—and we knew how to talk to them about participating in trials.

It was a meandering route, getting to this moment of discovery, but there was something beautiful about the way it had come about. People from all walks of life had sent in soil samples, and Brady had sifted through all of it, pulling out the most valuable molecule. Malacidins weren't the product of serendipity; it was a team effort, produced by and for average citizens, the way it should be. Perhaps Dr. Brady had found something that could help our patients. I sent a message to Tom—"Let's talk malacidins ASAP"—and set about the task of arranging a meeting.

Brady's work left me feeling invigorated, but that exuberance was about to change. It was time to inform Alicia that all of her diagnostic tests had come back negative. There was no evidence of Lyme disease, mold, or any other infection. She was still in pain and losing hope, and it was getting worse by the day. It would be a long walk to her room and an even longer conversation with her father. I was still searching for answers, and so were they.

But before I could explain things to them, I needed to check on an important patient: Gerard Jenkins.

CHAPTER 35

Angry Birds

THE APPOINTMENT WITH the hospital security guard was abbreviated. A snowstorm was headed toward Manhattan, and Gerard had only a half hour before he needed to get back to work. It was just enough time for me to collect the necessary information for the study: his rash had disappeared, his vital signs were stable, and, most importantly, he was back at work, pain-free. "No doctor's note," he said, shrugging his shoulders. Gerard was wearing a navy uniform and black leather gloves, and he was holding a walkie-talkie in his left hand. He looked much more robust than I had remembered, like someone who really could protect others.

"Sorry about that," I joked.

Hospital gowns have a way of sapping vitality, making patients appear far more fragile and debilitated than they actually feel. Sitting before me, Gerard looked like a different man. He passed the walkie-talkie from his left hand to his right and showed me his plastic badge. "They needed me back," he said. "It's all good." He flashed his gap-toothed grin and shook his head. "I actually kinda missed it." I ran through my list of questions, ensuring that Gerard hadn't developed an allergic reaction to dalba or needed further medical attention. "Nope," he said time and again as he stared at his leg, examining the area that had once been infected. "No problems. I'm good."

I put on a pair of disposable gloves and touched his skin, squeezing his calf in an attempt to elicit tenderness, but there was no response. Gerard was back to normal. "I needed to get back," he said. "No sense sittin' around all day, ya know?" He was pleased with the results and delighted that he would receive a $200 debit card at the completion of the study. When we were finished, Gerard's information was uploaded to a secure file and added to the mound of data I was submitting to the upcoming conference.

"Everything looks good," I said, handing him a sheet of paper summarizing my findings. "I'll call you in a few weeks."

His was a success story: Gerard was treated safely and efficiently and his speedy recovery had exceeded my expectations. Dalba had worked precisely as we had expected. He was back to work and back to his old self. I hoped my other patients would be so fortunate, but I knew that some of them would not be. I was due to give the drug to three more people later that day, all suffering from skin infections that had gotten worse after a course of oral antibiotics. "Stay warm," Gerard said, noticing that I was wearing a white coat but no overcoat. "Storm's coming." With a handshake, we parted ways. He went back to his job, and I went back to mine.

THE NEXT PATIENT on my dalba list was a twenty-three-year-old from the Bronx named Clara. She had sickle cell disease and had been in and out of my hospital nine times over the past year with what are known as pain crises. A single gene mutation—one that had protected her ancestors from malaria—caused Clara's red blood cells to become misshapen, forming tiny sickles that were sometimes so painful she couldn't breathe. This time around, she had come to the emergency room with a tender right leg and was admitted to the hospital for a presumed skin infection, one that was likely due to MRSA. When I entered her room, Clara was playing Angry Birds on her phone and *Dr. Phil* was on the television above her bed. She put down the game, muted the show, and waved me in. "What's up?" she asked.

"I'm Dr. McCarthy," I said as I grabbed a pair of gloves from a box on the wall and slid them on. I went through my standard introduction—I had whittled it down to two minutes—and pulled a plastic chair up to the foot of her bed. "I can go over the consent form if you're interested." She stared at me but didn't say anything. "I can address any questions," I said. "About the antibiotic or about anything else."

Clara shook her head and returned to Angry Birds.

"Would you like some time to think about it?" I asked. "Or I can come back later." I stood up and eased my way toward the door. "Or you can just tell me not to come back. It's totally fine." When I stepped away, I noticed a urinary catheter emerging from between her legs and made a note to talk to her doctor. Earlier in the year, Clara had developed a urinary tract infection with a bacterium called *Enterococcus faecalis*, and it had become resistant to most antibiotics. Scrolling through Clara's chart, I could see just how quickly it had mutated into a superbug. The catheter's plastic tubing would only make things worse, predisposing her to yet another infection. "I'll go," I said. "It was nice to meet you."

Clara looked up from her phone. "Anyone get sick taking it?"

"Great question." I sensed an opening and took two steps toward her. "A couple of people had rashes. Some nausea. I have a list of possible side effects I can share with you. If you're interested."

"What's in it for you?"

"Well," I said, "you get a new antibiotic, and for your time, we give you—"

"No," she said softly. "What's in it for *you*?"

"I get to study a new drug. I get to see if an antibiotic works. And I get to—"

She scratched her cheek. "You don't know if it works?"

"I get to see how *well* it works."

Clara looked at me askance. "What else?" Her eyes narrowed slightly.

"I'm running the study," I said, "so I'll get to present the results of the trial at a conference. Or several conferences. It could lead to more studies."

Clara's green eyes locked in on mine. "You gettin' paid?"

"I am," I said. When I'm nervous, I speak tend to speak too quickly, and I could feel myself racing through words and sentences. I opened the consent form and pointed to page four. "The company provides me with salary support."

"How much?"

I shifted in my shoes. It was the first time someone had asked me about financial relationships or potential conflicts of interest in more than a year of screening and interviewing patients. "About two percent of my salary."

"You own stock or whatever?"

"I do not." It was an awkward discussion, but the more Clara probed, the more relaxed I became. "Our hospital doesn't carry this drug. I'm the first to ever use it," I said. "A lot of people are looking to see how this trial turns out, and you can help with that. We both can. Or you can pass; feel free to say no."

Clara picked up her phone and started texting. "No, thank you," she said without looking up. "I'm not interested. You can go."

I put the consent form under my arm and took off my gloves. "You got it." Her words stung, but I tried not to show it. I stepped out of her room, went back to my office, and took off my white coat, replaying the exchange over and over. I try to learn something from every encounter, but sometimes the lesson isn't obvious. Sometimes the takeaway dawns on me weeks later, after I've seemingly forgotten about the patient. Then, out of nowhere, a random glance or an errant comment will send me tunneling back to somewhere else. The next time I see someone playing Angry Birds, I'll probably think of Clara.

I tried to imagine how our conversation might have gone better. I never cracked Clara's exterior, and my answers to her important questions did little to reassure her. Maybe the exchange had gone as well as it could have, considering the circumstances, but I was disappointed in how it ended. I was nothing more than an unnecessary intruder in her difficult life.

I put on "Take It to the Limit" to cheer myself up. Soon I was thinking about Randy Meisner, my kindred spirit. He was an anxious guy and occasionally felt uncomfortable performing his signature tune in front of crowds. It was the band's encore, and he was the only one who could reliably hit the high notes, but on some nights, Randy refused to sing it. That decision predictably enraged his Eagles bandmates, and it contributed to his departure from the group.

Randy Meisner on March 6, 1981, in Chicago, Illinois

Meisner had been surrounded by gifted musicians and lucked into one of the most popular musical acts of the twentieth century, but he wasn't able to meet their expectations. I suspected that he was intimidated by the geniuses around him, the ones who brought out the best in him, and he crumbled under the weight of it all. It was a bit of stretch, but I thought I knew how Randy felt.

I'd experienced the same self-doubt in the minor leagues every time I stepped onto the baseball field knowing that I was overmatched by an opponent. I experienced it when I entered Tom's office knowing that I hadn't produced the results he was counting on. And I continue to live with it when I step into the room of a patient like Alicia or Clara knowing that I'm not making a connection.

Eventually I turned off the music, put away my white coat, and decided to take a walk. I wandered west, toward Central Park, and spent the remainder of the day adrift in my thoughts, staring at pigeons and ponds and people eating hot dogs. Thinking about Clara. Was this trial about enlightened self-interest, I wondered, or something else?

Later in the day, I took a seat by the park's reservoir, just watching people go by, imagining the millions of molecules beneath their feet and mine, just waiting to be discovered. Sean Brady's team had used calcium as a starting point, but perhaps there were other, better ways to find drugs. The most powerful antifungal agent, amphotericin B, causes a rapid leakage of potassium and other ions, killing fungi and some parasites in the process. There were undoubtedly other molecules that could do something similar, but how could we find them? And who would pay for the fishing expedition?

As the first snowflakes started to blanket the city, I had a thought: there may be undiscovered antibiotics hidden in snow. Not the fresh powder dropping on the city, but the densely packed dry stuff near the Arctic. Tundra was known to harbor all kinds of interesting microbes—you wouldn't believe what's hiding in reindeer carcasses—and where there are bacteria, there are undoubtedly new drugs. I picked up my phone and called Sean Brady.

CHAPTER 36

Macaulay

ON WEDNESDAY MORNINGS in the spring, I teach an undergraduate seminar in medical ethics. The course is run by the Macaulay Honors College at City University of New York, and the students come from all over Manhattan and the outer boroughs. I teach with a gender studies expert named Elizabeth Reis, and we spend the semester confronting the most controversial issues the two of us can dream up. It was something I became interested in shortly after I started the dalba study, as I began to wrestle with the ethical challenges of informed consent, drug trials sponsored by pharmaceutical companies, and the blurry lines between treatment and research.

During the first few weeks of my study, I turned to seminal ethics texts—dense tomes such as *The Birth of Bioethics* and *Principles of Biomedical Ethics*—and was soon spending as much time with those works as with medical articles about antibiotics. The books exposed me to a new vocabulary and abstract concepts that didn't always jibe with my experiences on the hospital wards, and that inner conflict led me to the course.

Most of my students are premed, and many are from working-class backgrounds. Their parents are taxi drivers and cleaning ladies, they work at delis and day care centers, and the majority are first- or second-generation Americans. Some have been pushed toward medicine by ambitious parents, while others have found it on their own. We all show up each week to talk about how patient care can improve and to examine how the system can fail a patient like Alicia.

In my first lecture, I mention the two qualities I've found in almost all successful physicians: toughness and kindness. "A lot of people have one," I say, standing at a marker board, "but to combine them day after day is difficult, especially when you're exhausted and stressed-out." Then I mention my failings on both fronts, both as a trainee and as an attending physician. I conclude the first lecture with a bit of levity, likening medicine to parenting: "You're often woken up in the middle of the night," I tell the class, "to care for someone who doesn't particularly care about you, and might not do the same if the roles were reversed."

The ninety minutes we spend each Wednesday discussing these topics is one of the highlights of my week. (The other is receiving those predawn texts from Tom.) We cover seminal moments in medical ethics, including Tuskegee and Nuremberg and Henry Beecher's report, but most of our time focuses on the way medicine is changing, and what that means for doctors, patients, and other health care providers. Many of the topics are drawn from my own experiences, and in some cases, the students help me work through an issue that has left me stumped or feeling frustrated—like when a CEO cites an ethical requirement to jack up the price of an antibiotic. It often feels like group therapy. The students sometimes disagree with my approach, and with one another, but the conversation is always thoughtful. I try to remember that when I'm interacting with difficult personalities in the hospital or when I'm watching talking heads on cable news.

One week, I'll invite a colleague from psychiatry to discuss force-feeding anorexic patients; the next week, a hospitalist will talk about length of stay and the dual loyalty doctors have to their patients and their employers. Sometimes we visit my patients to hear about their experiences in the hospital. There are lots of questions and digressions, and frequent references to books from the syllabus. We also discuss informed consent and the challenges associated with conveying complicated concepts to vulnerable patients who may not have much health literacy. I use the case of Gerard Jenkins to illustrate the point: even if a patient provides consent, it might not be informed. How much effort must we put into that initial conversation, especially when the patient is willing to sign? And what happens when a volunteer asks for a small favor, such as a doctor's note? I remind them that there is still no framework to guide researchers in determining how to compensate volunteers for their time or, in some cases, their body fluids. For now, we just try to do what seems right, and the students remind me that there are many ways to interpret that.

After one lecture, a young woman asked a few follow-up questions about my trial and, more generally, about antibiotic discovery. This student was one of several who wore a hijab that semester, and she had just led the class through a discussion of ethical issues facing Muslim patients and doctors. She was one of the sharpest in the class, and she always pushed me to address my own biases when I confronted a thorny issue. Now she wanted to know why antibiotics were so hard to make. "Why wouldn't a hospital want to carry a new drug?" she asked. "Especially if it works?"

Soon we were chatting about Sean Brady, malacidins, venture capitalists, and the difficulty of bringing a molecule from the laboratory to patients. The student was intrigued by Brady's work and asked if anyone was doing something similar. "Why isn't everyone doing this?" she wanted to know. "It seems like the next big thing."

There was a look of genuine curiosity on her face, one that makes the class such a joy to teach, but I couldn't give her a satisfactory answer. "I'll try to find out," I said. "Maybe others are." I returned to my office and started to hunt.

I soon learned that one researcher, a biologist at Northeastern University named Kim Lewis, had indeed found something. Lewis's discovery was the result of a unique process, one that reminded me of Fleming's glass-blowing work a century earlier, which had fooled bacteria into growing in harsh conditions. Somehow, that led to the discovery of an antibiotic called teixobactin below a meadow in Maine. Just a few days after my student posed the question, I called Lewis to find out more.

CHAPTER 37

Searching

MAYBE IT WASN'T A MEADOW. Kim Lewis told me he doesn't actually know where his antibiotic came from. "The papers say a grassy field
in Maine," he said from his office at the Antimicrobial Discovery Center
at Northeastern University in Boston, "but we don't really know." In
fact, his team has chosen not to know, he said. "Someone we know was
on vacation and found it. It's an interesting story."

That tale begins more than a decade ago, when Kim Lewis and his
colleague Slava Epstein, a biologist who emigrated from Russia, were
contemplating a puzzle: Why don't most bacteria grow in petri dishes?
The two scientists batted the question back and forth, noting that more
than 99 percent of bacteria from soil cannot grow in regular culture
media in a laboratory, and decided it was worthy of investigation. It
would make sense to grow microbes in their natural environment, but
they weren't exactly sure how to replicate that in the confined environment of a research laboratory. Scientists had been confounded by this
problem ever since the petri dish was invented in 1887, and no one had
made much headway. They've been studying a thin slice of life, the 1
percent that could grow in a little dish, at the expense of all the rest. Like
two astronomers searching for a new galaxy, Lewis and Epstein wanted
to know what else was out there.

One day Lewis had an idea, or a fragment of one. "A very simple thing came to mind: we should use a dialysis bag, one that was designed for desalting proteins. We could fill one of the chambers with bacteria, place it in soil, and see what happens." A dialysis bag, used to remove toxins from the bloodstream, contains a semipermeable plastic membrane, and it allows chemicals to selectively pass from one side to another. By placing it back in soil, he could give bacteria natural nutrients not available in the laboratory, allowing them to thrive in an unnatural environment. If it worked, Lewis and his team would have access to thousands of undiscovered microbes and, more significantly, an untapped trove of new molecules.

The initial design didn't work, so they eventually created a newer, thinner version and in 2002 published the work in the journal *Science*. "No one paid much attention," Lewis said. "It was peculiar." (It was around the same time that Vincent Fischetti reported his groundbreaking lysin study, which was similarly ignored.)

Lewis and Epstein started a company with their technology, NovoBiotic Pharmaceuticals, and continued to improve their device. Epstein miniaturized the contraption in his laboratory, calling the updated thumb-sized invention the isolation chip, or iChip. "It was essentially a piece of plastic with a lot of little holes," Lewis explained, "something that you dip in a suspension of bacteria, cover with semipermeable membranes, and insert back in the environment the bacteria came from." The iChip was studded with minerals and nutrients, and it created a hospitable environment for thousands of previously undiscovered organisms to thrive. That piece of plastic ultimately led to the discovery of three dozen compounds, including teixobactin, the antibiotic from Maine.

Teixobactin was produced by a bacterium that no one had known about, one that previously couldn't grow in a laboratory. Lewis named it *Eleftheria terrae*. The organism presumably makes teixobactin to defend itself; the molecule is capable of destroying bugs such as MRSA, as well as *Mycobacterium tuberculosis*. Lewis and his team announced their findings in *Nature* in 2015 with the headline-grabbing title "A New

Antibiotic Kills Pathogens Without Detectable Resistance." What's notable about the manuscript is its simplicity: a head-spinning design is laid out in just a few paragraphs.

During our chat, Lewis spoke a bit about teixobactin, but mostly about bottlenecks: the areas where development slows down or grinds to a halt. This is where the biologist puts most of his energy and where he sees the greatest opportunity. His group recently won a five-year, $9 million grant from the National Institute of Allergy and Infectious Diseases to alleviate those bottlenecks and to improve the antibiotic pipeline. "The biggest problem is dereplication," he said, using a term I wasn't familiar with, "finding out whether your soil sample contains a new compound. You want to know if there's something potentially interesting. Is it new? Is it old? Is it junk?"

The vast majority of molecules that scientists find in dirt are unusable, either because they're toxic to humans or because they perform a function we already know about. Replicating old findings is a substantial waste of resources, and it's where these search parties often break down. "No one would've given me this grant to improve the current process," he said. "We're fundamentally changing it."

Lewis is familiar with Sean Brady's work, as well as the challenges of working with venture capitalists. He knows that investors are skittish, so he's continuing the hunt for drugs, using the iChip to identify ways to treat all kinds of diseases. His company has retrieved thousands of new compounds from soil to potentially treat everything from cancer to tuberculosis. "The reality is that considerable funds have already been put into drug development," he said, "but the bottleneck is in discovery. That's where we need to focus. I've got the only laboratory that's specifically addressing this issue." I felt a flicker of excitement, and the old urge to pick up a pipette.

After our conversation, I headed outside for lunch at one of the halal food carts across the street from my office. I stepped out of the aseptic hospital and into the crackling light of an early spring day, one of the first of the season, thinking about brown rice and hot sauce. It's always an odd feeling leaving a hospital, like walking out of a movie theater at midday, and my eyes struggled to adjust. I have a strange habit of sneez-ing when I emerge into bright light from darkness, and I had to grab a light pole to steady myself. I wiped my eyes with my sleeve and crossed the road. Tiny birds were chirping atop streetlights, and a patch of dan-delions had pushed through the soil on Sixty-Eighth Street, creating a dividing line of sorts between my hospital and Rockefeller University.

I bent down to take a closer look at the yellow petals, imagining what lay hidden in the dirt beneath them. Surrounded by concrete and sirens and skyscrapers, the soil was damp and full of life. I thought about the conversation with Kim Lewis and what I would tell my inquisitive ethics student—something about the mysterious field in Maine, certainly, and perhaps a bit about Lewis's technology, if I could ever wrap my head around it. *The bacteria had been tricked!*

I pressed my fingers into the ground surrounding the flowers, forget-ting momentarily that I was about to eat. Forgetting about all of the things that stress me out. In that moment, as I dug gently into the soil, I felt rejuvenated. The world was opening up to me in unexpected ways. I marveled at the simple reality that we were surrounded by undiscovered medicines. Microbes were engaged in biological warfare all around us, making new chemicals under our feet that could eventually end up sav-ing millions of lives. I was accustomed to thinking about the deadly in-fections that were coming for my patients, but now I could picture their cures, too. Just beneath the topsoil there were tiny molecules that could alleviate disease and stomp out epidemics. We just had to keep looking.

Chapter 38

Anna

Tom was eating a salad when the phone rang. We were due for a conference call with Remy's doctors in Germany, and I'd brought my takeout lunch to his office so I wouldn't miss the update. The team in Munich had been following Tom's radical recommendations for months, determined to cure the *Saprochaete clavata* infection in the young woman's spine. During the previous call, we learned that Remy was no longer on the verge of death, and we suspected the infection was no longer spreading. Containment was an important mark of progress: Remy and her family had started to think she might actually graduate from high school, albeit a year or two late.

"This is Dr. Levy," the voice said through the speakerphone. He had been coordinating Remy's care, ensuring that Tom was kept abreast of the tests and imaging, and was responsible for implementing any changes to her treatment. Levy told us that Remy had been moved to a rehabilitation center and that, while her fever was gone, she still had a dull ache in her spine. "I'm wondering if she needs a lumbar puncture," Levy said.

"What is her CRP?" Tom asked. It was short for C-reactive protein, a marker of inflammation, and he used it to assess clinical response in patients with spinal infections. It was an old-school test developed in the

1930s, and I hadn't been a big believer, but Tom had convinced me it was useful. I stirred my rice and chicken around the Styrofoam container, waiting for the answer.

"Trending down," the doctor said. "It's down to six."

"And how does Remy feel?"

"Better. She's walking and eager to get back to school, but there's still some pain."

"Walking! That's wonderful news."

Tom pulled up Remy's latest MRI on his desktop and waved me over. We could see that the infection was much smaller than before, and her leukemia appeared to be in remission. "We need to talk about end-points," he said. "How can we consolidate therapy and improve quality of life without risking progression of disease?" Remy was still on three intravenous medications to treat her infection, which limited her ability to leave the facility and return to normal life.

Tom pulled up Vivaldi's Concerto in G Minor as he did pharmaco-dynamic calculations in his head. "Let's decrease amphotericin B from five milligrams per kilogram per day to seven-point-five milligrams three times per week." The line was silent as the doctors scribbled down the recommendation. "That will make things a bit easier for her. We also need to address her pain." I did my own slower calculations as Tom outlined the sequence of steps that would gradually remove Remy's intravenous medications one by one until she was taking just a few pills that could be given at home. "The goal is to give Remy her life back," Tom said, "and to do so safely. There is no need for another lumbar puncture." He had never met Remy, but he was making the most important decisions of her young life.

"She'll be so happy to hear this!" Dr. Levy exclaimed. "She was terrified of having another spinal tap."

As for addressing the girl's pain, he offered them a brief tutorial on analgesics and the nonnarcotic options available.

"She has an appointment with a spinal surgeon next week," Levy said. "Do you mind if he calls you?"

"Of course not."

"My colleagues and I thank you, Dr. Walsh. Remy's parents thank you. We all thank you."

Tom hung up and turned to me. "Okay, what's next?" He clasped his hands together and looked up expectantly, but I wasn't quite ready to move on. We had several manuscripts to review and a new protocol to discuss, but those could wait.

"I've never asked you," I said. "What's your most memorable case? After all these years—forty-plus years—what's *the* one?" I knew he was managing at least a dozen patients remotely, all of them with strange infections, but I had never asked him if one was particularly striking. Or overwhelming. "Anything jump out?" I thought about how things had gone with Alicia and imagined how he might have handled her case.

Tom thought for a moment. "I suppose they're all memorable. In their own way."

I rolled my eyes. "Oh, come on, there has to be one."

He turned from me to his computer screen, staring at Remy's MRI for an uncomfortably long time, so long that I thought he'd moved on. "Well," he said. "There is one. Do you know Anna's story?"

IN FEBRUARY 2007, a six-year-old girl in Wisconsin named Anna was diagnosed with a rare form of cancer called high-risk acute lymphocytic leukemia. As the name suggests, it can be lethal, but there was a chance that her disease could be cured with a mix of chemotherapy and radiation. Anna was immediately taken out of school to begin treatment. She lost all of her hair and developed agonizing nausea, but the approach appeared to be working. One year later, though, and seemingly out of the blue, Anna lost feeling on the left side of her body. The doctors at Children's Hospital of Wisconsin discovered she had a large abscess in her brain that had caused a massive stroke. The chemotherapy had weakened the cancer, but it had also damaged her immune system, predisposing her to infection. After reviewing her brain scans, one doctor told Anna's father, "It's not a death sentence, but . . ."

A team of neurosurgeons took her to the operating room to remove the egg-sized pocket of pus, but when they opened Anna's scalp, they discovered several abscesses. The surgeons went about the delicate process of scooping out the infected material, which was due to a fungus called *Aspergillus*, but they couldn't get all of it without removing brain tissue. When she woke up, Anna still could not use her left leg or arm, and she had lost her vision. There was reason to believe the remaining infection would spread and kill her.

One of the doctors suggested something radical: complete removal of the right side of Anna's brain. The procedure is known as a hemispherectomy, and it's done in rare cases to prevent fungus from spreading to other organs. It was risky: if the girl survived the surgery, her personality would be altered, and she would lose the ability to have abstract thoughts, but it might be the only way to save her life. Anna's oncologist reached out to the one doctor they turned to when the chips were down and hope was all but lost.

"I couldn't believe they were going to remove half of this little girl's brain," Tom said to me as I finished off the last of the rice. "We had data indicating Anna's infection could be treated medically. Radical surgery was unnecessary. Absolutely unnecessary." He smiled as he recalled the initial conversation with Anna's father, Aleks. "He's a salt-of-the-earth type, kinda talks like he's a welder, but he's also got three degrees from MIT." Aleks was Tom's kind of guy: exceedingly sharp but didn't need to show it.

Walsh told the team in Wisconsin not to perform the hemispherectomy. Instead, he recommended stereotactic decompression—a form of microsurgery—and a combination of antifungal drugs that he'd been testing in his laboratory. The procedure would utilize a three-dimensional coordinate system to identify and remove the infectious material without disturbing the surrounding neurons. It was a last-ditch effort to save a child's brain, and no one knew if it would work. "After the surgery and the new medicine," Aleks would write later, "the miracles started to happen." Under Tom's guidance, Anna regained the ability to move her left leg and arm; not long after that, she started making a weekly phone call.

"Every Saturday morning," Tom said, "I would get this call from a six-year-old, giving me updates about her blood counts. I loved it." She continued to improve, and after forty days, Anna was discharged. "She walked out of the building," Tom said, "holding her father's hand."

It occurred to me that when Anna suffered her stroke, she was nearly the same age Tom had been when he lost his mother to cancer. It was just a coincidence, of course, but it seemed fitting. He was intervening to help families in a way that others couldn't, and I suspected it was partially because of his own childhood trauma.

"Anna's doing okay now?" I asked.

"She's about to graduate high school," he said, pulling up a picture on his computer. He dragged the image of Anna next to Remy's MRI scan. "Now she volunteers at the hospital where she was treated." Tom was beaming, and so was I. "Last time she and her father were in New York, we all met for pizza."

As he recounted the details of the method he used for treating her—which was a decade ahead of its time—I pulled up Aleks's blog, *Mission from the Heart: Appreciation and Support of Dr. Thomas J. Walsh— Sharing Stories of Hope and Healing.* It was full of testimonials from doctors and patients who had interacted with the man sitting across from me, and he had never mentioned it. As Tom told me about his pizza date with Anna, I read the words her father had written:

> Dr. Walsh's research discoveries have saved Anna's life. My wife and I are grateful beyond words for all the work he has done. Thank you Dr. Walsh for your dedication and passion. I sincerely hope that Dr. Walsh and his associates continue to discover and develop new medical advances that will ultimately help others to live longer or to survive a devastating infection.

CHAPTER 39

Reversals and Rewards

IT WAS A new experience for me to spend so much time in the hospital waiting. I was accustomed to hustling from one patient to the next, rolling one meeting into another, joining a conference call as I scarfed down lunch, but the reality of enrolling patients into a clinical trial is much different. It is *slow*. Most of the time is spent hovering around the action, like an extra on the set of a television show, waiting for your scene. Hoping to hear "Action!"

During the dalba study, I logged hundreds of hours at the bedside, waiting quietly as men and women tried to make up their minds about a drug they knew relatively little about. I often found myself staring at a clock, watching the minutes tick by in silence, wondering what the person across from me was thinking. *Will she or won't she?* The irony of this role reversal was not lost on me. For once, the doctor was waiting on the patient.

A recurring pattern played out throughout the study: just as I thought a patient was about to sign the consent form, he or she would hand it back and ask to think it over. I would offer to circle back in an hour or two and mention a few useful websites about antibiotics. "Take your time," I would say as I closed the door. Upon my return, I would discover that the patient was no closer to making a decision and was now waiting for a family member or friend to contribute his or her opinion. "No problem," I would say, "take as much time as you need."

There was no silver lining or nugget of wisdom to be gleaned from these moments. They were just a mundane yet necessary aspect of clinical research that no one talks about. For some, the extra time and information helped to clarify a complicated decision; for others, it simply added to the confusion. When I returned to a patient's room, I could never predict if I was about to enroll a volunteer or not. And in a way, that uncertainty provided its own thrill.

One of the final patients I enrolled was an elementary school teacher named Jennifer. By the time I met her, I had screened thousands of patients and had spoken to hundreds about informed consent. No two conversations had been the same, and in every single one, I had said or heard something that surprised me. Before I entered Jennifer's room, I had been evaluating a stoic man with three small tears tattooed next to his left eye and a large rash on his thigh. I was concerned that the ink represented prior misdeeds—one urban legend suggested the teardrop represented attempted murder or a lengthy prison sentence—and I was conflicted about his participation. But we hadn't discussed any criminal activity, and he wanted to enroll, and withholding dalba seemed wrong. "Shouldn't be a problem," I told him. "Let me just confirm that you're eligible." He was.

In Jennifer's room, I was met by a beam of early-morning sunlight and raised my right hand to shield my eyes. It occurred to me that it was a silly but appropriate symbol: my years-long project was finally coming to an end, and this was the light at the end of the tunnel. I closed the door, pulled out my business card, and introduced myself. Jennifer taught fourth graders in a town near mine in Westchester County. There was nothing particularly remarkable about her case: she had a painful red rash on her arm that hadn't improved with oral antibiotics and had been sent to the emergency room by her primary care doctor. "I just want to get back to work," she said as I examined her skin. "I don't know what happened. Maybe I had a bug bite?"

She had a MRSA infection, and I presented her with two options:

hospitalization for treatment with intravenous vancomycin or a single dose of dalba. The former would require several days in the hospital; the latter would have her home in a matter of hours. What I had discovered over the course of the trial is that many patients take comfort in receiving daily treatment and aren't necessarily happy about being discharged without a bottle of antibiotics. "One dose?" some would ask. "That's it?"

Jennifer adjusted her glasses when I handed her the consent form. "What I'm wondering," she said as she scanned the pages, "is what I can do to protect my class." She looked at the ruby welt on her forearm. "I don't want the students to get this." There was something about that comment that pierced me. She cared about her own well-being, of course, but there was a classroom full of children that was also on her mind. I pictured a chattering class of students welcoming their teacher back after an unexpected absence. Maybe there would be a cake. "Let's talk about that," I said, taking a seat at the edge of her bed. "I've got a few tips."

Jennifer took notes as I spoke, documenting carefully how she could protect her students from skin infections. We spoke about bug bites, spider bites, dog bites, Lyme disease, and even a bit about Fleming and Gerhard Domagk, the fortuitous grenadier. "Oh," I said, tapping my thigh, "you've got to tell them how nystatin was discovered. It's a wild story. In the 1950s, these two scientists in New York, Elizabeth Hazen and Rachel Brown, started mailing dirt to each other . . ." Jennifer wrote *nystatin* on her notepad and underlined it.

"I could teach the kids about this," she said. Her look of concern faded. "After I'm better. Something about how infections happen and why antibiotics work. The kids will love it!"

This was the ebb and flow of clinical research: long periods of tedium punctuated by a flurry of excitement. These were the transcendent moments that would outlive the brief exchange between a doctor and a research subject. These were the moments that made the trial so rewarding. "Maybe I could help," I said. "I think about this stuff a lot."

An hour later, after the consent form had been signed and the final drops of dalba had been infused into her bloodstream, Jennifer gathered her belongings, stuffed them into a red duffel bag, and headed home to Westchester. She was back with her students the following day.

Two weeks later, Jennifer arrived in my office with a beaming grin and a card for me. "From my class," she said as she took a seat and showed me the area where the rash had been. It was completely healed, and she was back to life as usual. Her class had just learned about antibiotics, she said, and infections. I opened the card and found a dozen names scribbled around a large handwritten message:

Thank You Doctor Macarthe!

CHAPTER 40

Help Wanted

"Before I forget," Tom said. "I want to tell you something." It was the end of a long day, and we had just finished drawing up plans for our next study. A rival company had developed an antibiotic similar to dalba, and we were excited to see if it worked, too. We wanted to broaden our focus, to examine bones and joints that been infected with drug-resistant bacteria. There was a chill in the air that evening, and the elm and honey locust trees around the city had just turned green. "I'm going to nominate you for something."

"Oh yeah?" I took a bite of a turkey burger, one of a dozen I consumed every week at the hospital. While I was standing in line to buy it, I bumped into Erwin, the playful medical student from Nebraska. He was now a doctor, with a long white coat and a world-weary look on his face. When he saw me, he smiled and said, "Doctor." I glanced at his chest, hoping I might see a nipple ring poking through, but there was just a gaggle of pens and folded papers peeking out of his breast pocket.

"So, this nomination," Tom went on. "I think you'll be a strong contender."

"I'm listening." He pointed at his computer. Next to a picture of Tom standing in front of a bookshelf, it read: "Thomas J. Walsh Young

Investigator Award." The prize was sponsored by the Medical Mycology Society of the Americas and designed to develop the next generation of physician-investigators. I shook my head and wisecracked, "I hear that guy's a piece of work."

"Me too."

He scrolled down as I read the eligibility criteria. "You'll need to draft an acceptance speech," he said, half joking.

I looked down at my notebook. It was filled with hundreds of pages of scientific projects, ideas, and random observations about Tom. I tried to think of a scene from *The Karate Kid* that might capture our dynamic. To many, Tom was an enigma, a man constantly on the move, but I had seen him in the quiet moments, agonizing over the management of a child on another continent, absorbing the pain of others, getting drugs to those most in need. Carrying on the mission.

I had appointed myself his Boswell, spending a decade collecting his thoughts, hoping I might absorb some of his wisdom and decisiveness. I knew so much about this man: his favorite words, his favorite symphonies, his thoughts on arcane grammatical rules. I even knew how fast he could run in high school. I knew more about Tom Walsh than I did about many members of my own extended family.

"I suspect you're right." I removed a pen from my white coat and held it between us for just a moment. "But where?" I said. "Where should I begin?"

Tom dismissed the question and put down his pen. Then he squeezed me lightly on the shoulder and ran through the checklist that concluded every meeting. "How are your patients?" he asked. "And your papers, and your family?"

I gave him a brief update about the first two, but I needed his advice about the third. "One of my family members isn't doing so well," I said, once again blurring the lines between professional and personal. I hesitated momentarily but pressed ahead. "And I'm not sure what to do. It's my father-in-law, Bill."

Tom's face fell. "What happened?" he asked. "Tell me about him."

————

BILL MORRIS LIVED a New York life. My father-in-law was born shortly after World War II on the northern tip of Manhattan, in Washington Heights, back when it was mostly Irish and Italian, to a police officer and a homemaker who was a talented pianist. The blue-eyed boy was a rambunctious kid but a good student and a talented athlete, and he tested into Stuyvesant, the city's premier public high school. From there it was on to Hunter College, one of Manhattan's best public universities, along with a string of odd jobs around the city—security guard at Shea Stadium, gardener at Rockefeller Center—and a long career as a New York public school gym teacher, coach, and varsity softball umpire.

Bill married another teacher, Harrel, and together they had two children: Jonathan and Heather, who became my wife. She and I met during our first year of medical school, and we bonded quickly over our similar backgrounds. Our fathers are both teachers, and we spent much of our suburban childhoods on ball fields, learning from them. From the moment I met Bill on a snowy night in Boston, I marveled at the bond he shared with his daughter: Heather and her father finished each other's sentences, and she felt at ease when he was around. They laughed at their shared neuroses and reveled in their diverging opinions about current events—a repeated note of contention between them was Barack Obama's claim "If you like your doctor, you can keep your doctor."

Not long after I met him, Bill told me he wanted to be buried with just one thing: a letter Heather wrote about him when she was a senior in high school. It was just a few handwritten pages, but it meant everything to him. It reminded me of the nine-page letter Tom Walsh wrote about his daughter, and of the tattered binder that Alicia's father carried from one hospital to the next.

Bill didn't like to travel, but Heather convinced him to take her stepmother, Susan, to Tuscany for a vacation. It was a long trip, nearly two weeks, and Bill was a wreck before the flight. He was anxious about deviating from his routine and almost canceled. But Heather insisted it would be good for him.

The trip was a disaster. As they bounced across Italian vineyards, Bill developed a swollen ankle, along with diarrhea, dark urine, nausea, a persistent cough, and an odd-looking rash on his chest. When he returned to New York, he went to his primary care doctor—the same physician who treats everyone in their family—and discovered that his liver tests were off the chart. Heather knew that Bill had pancreatic cancer before anyone else did.

Before I had time to digest the diagnosis, she leapt into action, calling every oncologist she could find to identify the top pancreatic specialist in the city. Heather knew the data—most patients were dead within six months—but she was determined to save him. When Tom asked me how my family was doing, I was grappling with Bill's diagnosis, grasping for some way to prolong his life. A cure seemed unlikely, but maybe there was something we could do.

Heather found an oncologist at her own hospital, and we soon learned that Bill's form of cancer was deemed "borderline resectable"—meaning that surgery, known as a Whipple procedure, might work or it might not. The tumor was abutting a major artery, and it was unclear if resection was safe or even possible. Heather and I had both cared for patients with pancreatic cancer, and very few lived more than a year. We chose not to share this with Bill or the rest of our family, but they knew.

Before the surgery, my father-in-law needed several months of chemotherapy to shrink the tumor. He enrolled in a multicenter clinical trial and was soon signing consent forms not altogether different from the papers I carried around in my white coat. The trial called for a combination regimen called FOLFIRINOX. (*FOL* stood for folinic acid; F represented fluorouracil; *IRIN* was irinotecan; and *OX* was oxaliplatin.) If the tumor spread from his pancreas, he was a dead man; if the chemo could contain or shrink the tumor, he might have a shot. "So this is the thing," Heather would say to me in idle moments. "This is what will kill my father."

After two months of intensive chemo, the tumor had shrunk by

10 percent, and surgery seemed like a possibility. But the treatment was also hurting Bill's immune system, and we knew it was only a matter of time before some bacteria or fungus worked its way under his skin, causing an infection that could derail everything.

A few weeks after the scan, it happened. Bill developed body aches and back pain the day after I injected him in my kitchen with a dose of Neulasta, a medication meant to boost his immune system. Over the next two days, the pain worsened, and he went to an emergency room near his house in Putnam County, New York, where he was diagnosed with a staph infection in his blood. The diagnosis came as a bit of a surprise: Bill didn't have a fever and hadn't looked all that ill. He was started on vancomycin and then transitioned to oxacillin, a cousin of penicillin, to eradicate the infection.

It didn't clear. Despite powerful doses of the antibiotics—he received oxacillin every four hours—staph remained in his blood, and the back pain worsened. Heather was at her father's side for all of it, coordinating his care with a series of increasingly specialized consultants. I left dinner on Christmas Eve of 2017 to evaluate him myself, putting him through an exhaustive neurologic examination to determine if the infection had reached his heart or spine. (I didn't think it had.) In that hospital room, I felt the indignity of disease: the man who had been so virile just a few weeks earlier could no longer stand and had become incontinent. It took three people to move Bill from his bed to a chair. He looked like a different person, sapped of his vitality the way Gerard Jenkins had been. Heather said he looked like a ghost of her father. The thought almost made her vomit.

After I told Tom, he checked in every day and offered to consult on the case. It was surreal to hear him talking about my father-in-law, recommending imaging studies and devising treatment strategies, just like he had for all the others. I could occasionally hear him scribbling on a notepad as we spoke on the phone, catching every last detail before beginning his calculations. Tom's involvement was comforting, but it was also a reminder of how perilous things had become.

During one of her late-night visits to the hospital, Heather noticed that her father's bladder was swollen and he was having trouble urinating, and, contrary to my assessment, she believed staph had spread to his spine. She suspected that the infection was expanding and that antibiotics weren't working. When I reexamined him, I discovered that she was right. Bill was in agony and no amount of morphine or Dilaudid could help him. Her chance observation set in motion a cascade of events, culminating in his transfer to Columbia-Presbyterian, where an MRI revealed a large spinal abscess, and he was whisked away for emergent surgery.

In the operating room, just a few blocks from where he grew up, a team of neurosurgeons went about the task of physically removing the infected material from his body. It was in that moment, as we waited anxiously for an update, that I felt the supreme limitations of antibiotics. Here we were, on the cusp of so much scientific progress, but what Bill needed was for someone to go in and just cut out the damn infection. The advances made by Fleming and Domagk and Walsh and all the others were worthless. Novel treatments in the future might be useful—one day we might use lysins to treat Bill's infection—but when he needed a cure, it was a team of surgeons with knives who saved him.

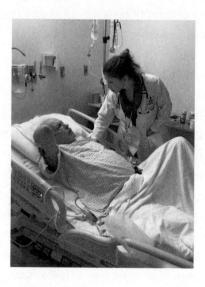

Heather with Bill after neurosurgery

Heather and I had a number of difficult conversations after that. I had argued that spinal surgery wasn't necessary and that the antibiotics should be given a chance to work. But they *hadn't* worked. I stood by my assessment—he had high-grade bacteremia and was a poor surgical candidate when I first examined him—and she stood by hers. In those tense moments, we spoke to each other like two physicians with differing opinions, not as husband and wife. I ultimately came to realize that I was wrong, but I had trouble admitting it.

"I have one question," I said to Bill shortly after spinal surgery. He was weak and groggy, slowly waking from the haze of anesthesia, and he had just contracted a rhinovirus, which gave him a runny nose. He was now in a private room overlooking the Hudson, to prevent spreading the viral infection to others, and the television was showing NFL football highlights on *SportsCenter*. We both looked out at the murky river.

"Did you like your doctor," I asked, "the one who operated on you?"

He furrowed his brow. It was a straightforward question, an easy way for me to assess his mental status, but Bill was having trouble responding. After a moment, he said softly, "Yes." The procedure had been performed by Alexander Tuchman, a thirty-five-year-old neurosurgeon, who had spent two and a half hours scooping pus out of Bill's spine.

"Because, Bill, you know, he's very good, and . . . if you like your doctor, you can—"

He flashed a slight smile. "I can keep my doctor!"

My father-in-law spent the next month in the hospital, gradually regaining the ability to walk and urinate on his own. After four weeks of physical therapy, he was discharged with a continuous infusion of oxacillin that was administered through a catheter in his arm. Although the antibiotic hadn't helped before surgery, we hoped it would prevent any further spread of the infection. However, this all meant that chemotherapy was on hold, and the pancreatic tumor was possibly growing.

After a few weeks of physical therapy, Bill's pain improved, and he went back to being the guy I'd known: the one who lit up in the presence of his grandchildren, the one who quietly gave me a hard time for not saving more for retirement. Two months after neurosurgery, his symptoms had almost completely resolved, and the infection became an afterthought. The antibiotic was stopped, the catheter was removed, and we began to use a word that had been verboten for months: *progress*.

Bill reconnected with his oncologist and starting planning the next steps in his cancer treatment, which would now include radiation therapy before the Whipple surgery. It was a massive procedure—the case I had scrubbed in for had lasted eight hours—and there was no guarantee it would work. But it was his only shot. Hovering over the plans was the specter of infection; another fever could further delay the operation, giving the tumor a chance to metastasize, transforming from borderline resectable to unresectable.

Not long after stopping oxacillin, Bill told us about a dull ache in his back. It started when he was watching television, and it stung when he crawled into bed. Two days later, as we weighed his options—did he need to go back to the ER?—Bill developed a foot drop. He could no longer flex or extend his ankle, and had to swing his leg when he walked. Heather and I scrambled up to his house to examine him and drew the same conclusion: either the infection had returned or the tumor had spread to his spine. "If it's the tumor," she said as we drove home, "it's game over." Her eyes welled up, and so did mine.

Heather stayed up late that night researching the treatment options— a different antibiotic, more surgery, something experimental—while reaching out to all of Bill's doctors. There was yet another possibility: Bill could have a different infection, one caused by something other than staph. More testing was in order, and we knew that his chances of survival were plummeting. While he was fighting off the infection, the pancreatic tumor was growing.

On more than one occasion, I dreamed of asking Vince Fischetti for some lysin to give to Bill. I kept thinking of his videos of exploding bacteria. But it was all fantasy: lysin hadn't been approved by the FDA, and Bill didn't qualify for an experimental trial. There was simply no way to give it to him, so we turned to his doctors at Columbia-Presbyterian. His neurosurgeon and infectious diseases specialist were equally concerned. His inflammatory markers, including the CRP test that Tom found so useful, were elevated, and his MRI was a mess. One by one, his physicians hinted at what had become apparent: Bill needed more spinal surgery to scoop out the remaining infection. "This is murky territory," his neurosurgeon wrote to Heather, "to say the least."

After two weeks of deliberation, Bill made the decision to forgo another spinal surgery. He just didn't think he could handle it *and* the Whipple. It was a lot to ask of anyone, especially a man in his seventies, so he opted to live with the numbness and weakness, altering his gait to accommodate the foot drop.

After the initial staph infection, Bill's doctors deviated from the initial plan and added radiation therapy to his treatment regimen in the hope that it might help him. I asked Tom what he thought of this wrinkle in Bill's treatment, hoping he would approve. "Does radiation make sense for pancreatic cancer?" I asked. Walsh was one of the few people in the world who was an expert in both oncology and infectious diseases, and if anyone would understand Bill's predicament, it was him. I outlined the plan—radiation, surgery, and then more chemo—and held my breath.

"Radiation may affect the surgical bed," he said. "It may make the Whipple more difficult." I valued his honesty, even if it was tough to swallow. "It may help," he added, "but it's risky." He put his hand on my shoulder, and then we stared at each other for what might have been ten seconds or two minutes.

"This is hard," I said. "It's just . . . I don't know. It's hard."

AS BILL WAITED, Heather trained. My wife is not a natural long-distance runner, but she signed up for a half marathon because she wanted to go through something long and difficult with her father. That's always been her style: if he was going to endure pain, she would, too. If she had been allowed to take the chemotherapy with Bill, she would have.

Heather joined Project Purple, a nonprofit that raises money for pancreatic cancer research, and she spread the word that her dad was battling cancer. She set up a webpage, wrote a brief testimonial, and spent the next five months preparing for the roughly thirteen-mile race. She ran late at night, after she'd finished seeing patients for the day and had put our kids to bed. At age thirty-six, she learned how to run in the dark.

The New York City half marathon took place on a bright, frigid Sunday in March 2018, just a few weeks before Bill's pancreatic surgery. A small group of friends and family gathered on the edge of Central Park to cheer her on, while Bill stayed at home, recovering from radiation and physical therapy. (I offered to fill Heather's iPod with Eagles songs; she declined politely.) We had a mobile app to chart my wife's progress through Brooklyn and Manhattan, and just before the race started, when the temperature finally rose about thirty degrees, I texted two words of guidance: "Go slow."

She didn't heed my advice. Heather completed the race in just under two and a half hours, much faster than any of us expected. At the after-party at a restaurant on Fifty-Fifth Street, the CEO of Project Purple, Dino Verrelli, a man who had lost his father to pancreatic cancer in 2011, singled out Heather: among sixty-seven runners, she had been the top fund-raiser. "Heather's dad is being treated at Columbia," he said to the crowd. "She told me that he looks good to have the Whipple in a few weeks, which is the best news I've heard today." The group cheered and rang cowbells as a kid in a purple unicorn outfit handed Heather a trophy.

There was a mix of joy and sadness in the room that day, recognizing just how many lives had been touched by the same disease, and more than a few tears were shed at our table. Tom Steitz, the Nobel laureate who first piqued my interest in antibiotics, would receive a diagnosis of pancreatic cancer a few months after Bill. Jon Steitz, my former baseball teammate, would soon run in a fund-raiser to support his dad, too. (Tom Steitz passed away in October 2018.)

At the half marathon reception, I learned that Project Purple doesn't just support research, it also has a financial aid program to help patients cover the costs associated with pancreatic cancer. The program laid bare a troubling reality: some are able to cover their medical expenses only through donations. Many would file for bankruptcy during treatment. As I listened to the presentation, I thought of the bill Tom Walsh's father received after his wife died, and the doctors who refused payment.

When we arrived home that afternoon, my son, Nathan, was waiting for us in the doorway. He told Heather that he was proud of her and, more importantly, that he'd seen something new outside of our house, something that no had ever seen there before. He led us a few steps from the front door and motioned toward the ground, pointing at a short row of crocuses. "Look," he said, "purple flowers." He crouched down, grabbed one by the stem, and ripped it from the soil. My son placed the little crocus in my hand and squeezed it, indicating that it was now mine. But I didn't look at it. My eyes were still focused on the tiny patch of dirt where the flower had been, imagining what microscopic miracles might lie beneath.

Epilogue

ONE AFTERNOON NOT LONG AGO, I sat down with Tom to extend an invitation. I wanted him to speak to my ethics students, and although I wasn't sure what the topic should be, I figured he could think of something. So much of his work had teetered on the edge of what was thought possible, placing doctors and patients in unfamiliar circumstances, advancing science and medicine faster than ethicists could possibly keep up. Many of the lives he saved for free; he never charged Anna or Remy or any of the other countless patients for his time and expertise. It had simply been part of the mission. Whatever Tom Walsh decided to talk about, I knew it would benefit my class. Or any class, really.

I also wanted to give him an update about Bill. When the surgeons opened up my father-in-law to perform the Whipple procedure, they discovered that his pancreatic tumor had spread to a large vessel in the abdomen known as the portal vein, which carries blood to the liver. In order to remove the malignancy, the team at Columbia-Presbyterian, led by John Chabot, had to remove both the cancerous mass and the fragile vessel, replacing it with a strip of the left jugular vein from Bill's neck. It was an extraordinary procedure, one that few surgeons could pull off safely—but it worked. The Whipple took about six hours, and when my father-in-law arrived in the ICU for postoperative monitoring,

Heather was waiting for him. "He made it," she said to me, fighting back tears.

Bill had survived the Whipple, but he was not yet free from danger. He needed more sessions of chemotherapy to ensure that all of the cancer cells in his body had been destroyed, and there was still staph bacteria circulating in his spine. Chemo would put him at risk for another life-threatening infection, and he would need antibiotics to keep the bacteria at bay. But which ones? And for how long? These were questions I couldn't answer, but perhaps Tom could. It was still touch and go, but Heather's quick thinking had saved her father's life.

I knew that talking about Bill's case could go on for hours, so I wanted to get a few other topics out of the way first. I began our meeting by mentioning something to Tom that I had just read: the Access to Medicine Foundation had created an index that ranked pharmaceutical companies according to their efforts to combat superbugs, and GlaxoSmithKline was at the top. A variety of metrics were used, including responsible manufacturing, research and development, and antibiotic stewardship, and the results were revealed at the World Economic Forum in Davos, Switzerland. GSK earned the distinction because the company had done something unusual: it started rewarding employees for helping doctors prescribe antibiotics *appropriately* rather than for simply meeting sales quotas. It was a radical departure from the status quo and signaled a change in the way the industry planned to contribute to public health. The award was the culmination of a decades-long comeback, indicating that GSK had fully recovered from the disastrous antibiotic hunt of the late 1990s.

A day after the rankings were released, an international consortium of academics and representatives from the pharmaceutical industry released a new economic model to spur drug development, calling for a $1 billion market entry reward for antibiotics that meet prespecified criteria addressing urgent public health needs. The report, funded by the European Medicines Initiative, a public-private partnership that aims to improve drug development, was novel because it disentangled profit from sales and, if fully implemented, could bring nearly two dozen antibiotics to market over the next three decades. When I showed the model to Tom, he was delighted. "This is right on the mark," he said. His face lit up as he reviewed the report. I took it as a sign of a small shift in corporate mentality, away from the idea that antibiotics were just like any other product, and that there was an ethical mandate to not simply maximize profits.

Eventually our conversation turned to malacidins and teixobactin and the tantalizing new methods for finding antibiotics. Drug discovery was changing, and we were both encouraged by the pace. I finally had a new way to treat Jackson, one that would safely target the unique features of his infection without destroying his vital organs. We were also pleased by the results of our dalba study, which proved that an expensive and dynamic antibiotic could be introduced into a major metropolitan academic medical center and could potentially pay for itself. I had grown accustomed to patients hugging me at follow-up visits, grateful that their painful infections had been cured by a drug they'd never ever heard of before.

The deeper I had gotten into the study, the more I came to appreciate a prior nemesis: the institutional review board. The IRB had been designed to protect patients, and that's exactly what it had done. Many of the men and women decided to enroll in my trial only after they knew it had been approved by the hospital. A blessing from the IRB meant something, and it was my fault that there had been such a delay in implementing the protocol. When I went back and looked at the early drafts, I discovered that the wording had been confusing, and it was the right call to slow things down. That delay gave me time to think through everything, to ensure it was properly executed. I had been frustrated in the moment, but the IRB had done its job, and my subsequent trials sailed through the approval process. I even became friends with a member who serves on the committee. The delay was my fault and mine alone.

I was also able to appreciate the difficult work that must be done by the Food and Drug Administration. Protecting patients is a big responsibility, and the agency must balance a number of competing interests from manufacturers, doctors, drug developers, and, of course, patients. We all share a common goal—to bring effective new drugs to those who need them most—and occasionally we'll differ on the best way to get there. I had learned a lot studying dalba: about antibiotics, ethical research, and the economic realities of medicine. I also learned about my own limitations.

One of the unforeseen challenges of the study was that many of the potential volunteers were homeless. They sleep upright on park benches or in subway cars, and the awkward positioning causes blood to pool in their legs. Over time their ankles swell, and occasionally the overlying skin bursts, allowing bacteria to enter. Once that protective barrier has been compromised, an infection develops, and it can tunnel into surrounding muscle, bone, and blood vessels. Through the trial, I saw that many of these patients are skeptical of doctors, especially guys like me who carry consent forms and speak of experimental medications. Building trust is never easy in medicine—Alicia packed up her things and left the hospital before I could provide her with test results—and I struggled to establish a rapport with many of the men and women who qualified for my trial. They entered the emergency room conflicted, and I wasn't the type of help they were searching for.

These were the patients I thought would benefit the most from the study—the ones who didn't have insurance or a pharmacy or a way of storing medications—but I struggled to convey that. I wanted to be the one to give them an expensive and powerful drug, but the looks I received day after day served as a constant reminder of my inability to connect meaningfully. A number of these patients told me that they were simply looking for a warm place to sleep, and the hospital seemed like a pretty good option. The idea of enrolling in a trial to facilitate a quick discharge back to the street was the last thing that they wanted. At first, I had trouble accepting that. My enthusiasm for dalba and the study impaired my ability to empathize with people whose lives were very different from my own.

I still brought a ballplayer's mentality to medicine, eager to win approval in the form of consent, but I have since learned to take the long view in clinical research. The nourishment of slow and steady success is far more meaningful than a quick victory. It was only one trial, of course, but with dalba, I had proof of principle—solid evidence that my ideas had practical applications—and I felt like I was making a statement. Antibiotic trials could be conducted at our large urban teaching hospital with a bare-bones team, one that worked closely together and believed in the product. Our study had shown the viability of an alternative treatment for aggressive skin infections, and we hoped others would take note. We had earned our Michelin star.

Dalba had captured the imagination of nurses, doctors, case managers, social workers, and countless patients who had been confounded by superbugs. Physicians from all over the hospital were asking me how they could get their hands on it. In many of those conversations, I was compelled to urge caution. The drug wasn't a cure-all, and some patients experienced adverse reactions to it. If we used it indiscriminately, bacteria would develop resistance. I tried to heed Brad Spellberg's warning—bacteria use antibiotics judiciously; humans do not—but the story of my trial spread quickly to other health care centers. I mentioned to Heather that perhaps her transplant center might find a use for dalba, too.

Physicians at other hospitals started using the drug to treat all sorts of life-threatening conditions, including bone, heart, and bloodstream infections, and it had contributed to an improvement in the delivery of care across the country. Patients were no longer waiting for days in the ER just to find a hospital bed, and health-care–acquired infections at some facilities started to dip. It was a joy to walk into a relatively empty emergency room with doctors and nurses calmly awaiting the next patient. I was receiving calls from doctors who were excited to replicate our work. Just down the street from me, researchers at New York University Medical Center found that dalba decreased emergency room length of stay from twenty-six hours to just five. The finding had ramifications for everyone, even those who will never develop a drug-resistant infection.

Soren Gillickson, the young man with the intravenous drug habit, had been excluded from my trial over fears that he could have a bloodstream infection, one that might not respond to an antibiotic such as dalba. But after my study was over, doctors had shown that the drug could work in that condition, too, and, remarkably, some could be treated as outpatients. This had been unthinkable just a few years ago, but dalba was opening up unexpected possibilities, providing new options for patients who were eager to avoid hospitalization and for doctors who were tasked with allocating scarce resources.

The medical system had failed Soren, exposing him to whopping doses of narcotics after his surgery, transforming him from an ambitious young man into a reclusive opiate addict. We will never fully atone for that tragedy, but with dalba, we're providing an important therapy that might stop others from following his unfortunate path. For some, avoiding hospitalization might prevent exposure to opiates and the descent into addiction. For countless others—people like Ruth and George and Erwin and Gerard and Jennifer—dalba represents a way of returning to normal life, removed from stretchers and X-rays and blood draws—away from all of the unforeseen perils of the modern hospital. It's a mechanism for deflecting danger.

Advances in health care often occur at the margins, where innovation seems impossible, impractical, or unnecessary. Most physicians aren't particularly interested in studying bacterial skin infections, and at first blush, the implications of our study might be difficult to appreciate. But the data are impossible to ignore: the clinical trial had shone a light on an area of medicine where most people weren't looking, where lives could be improved, and where medical care can be streamlined. I thought of Ruth and her accidental fall. How many of those had we avoided by challenging the status quo?

More than twenty million people worldwide develop a skin infection every year, and nearly twenty thousand of them will die. Dalba represents a new way of treating them, and a dramatic shift in the allocation of health care dollars. I knew Tom shared my enthusiasm and could see possibilities far beyond the ones I had already identified. He was meant to do this stuff—I had spent nearly every day over the past decade being reminded of that—and despite my impatience and occasional cynicism, I finally believed that I was meant to do this stuff, too.

This model, I suspect, will eventually weave itself into drug development. I learn about new medications all the time, but they're invariably expensive, and there's always disagreement about how to utilize them best. Hospitals are conservative by nature—administrators, thankfully, try to keep disruptions to a minimum—and the decision to try a drug, even a fabulous one, has to make financial sense. There's a reason people are hesitant to invest in antibiotics, but our study might give them a nudge.

The day after my dalba study ended, a pharmaceutical rep called to gauge my interest in investigating a new influenza drug. "It's given as a single dose," he said, "just like dalba." He thought I could model the influenza study on the work that had been done with cellulitis. It would be another annus mirabilis.

JUST BEFORE I brought up lingering questions surrounding Bill's case, I mentioned to Tom that he was going to be a focus of the book I was writing and suggested we set up a formal interview to get his thoughts on where the field of superbug research was heading. "I'll probably need an entire chapter," he said. "You know, once I get going, it's hard to stop."

"I hadn't noticed."

"Things change so quickly," he said, offering a note of caution. "You may have to rewrite the book once you're done."

"I hope not. You think so?"

"Are you familiar with Heraclitus?" he asked. I shook my head as Tom leaned closer to me. "He's credited with saying that no man steps in the same river twice." He pulled out a notepad and retrieved a textbook from his shelf. "I'll be interested to see how you define *superbugs*," he said. "Is influenza a superbug? Or HIV? Are you referring to antibiotic-resistant bacteria?"

The question caused my mind to race. "I'm sure I haven't captured everything," I said, trying to sound calm. I thought about all of the infections I had seen over the years and the diverse group who had volunteered for my study. I thought about ombré-dyed hair and a boy in Superman pajamas.

"It would be impossible to do so." He patted me on the back and started flipping through his notes on CRISPR and bacteriophages. Soon Tom was drawing a large molecule on a yellow sheet of paper, carefully arranging the carbon atoms in a neat row. "I have an idea," he said, pointing at the picture. "Do you have a moment?"

In our decade of collaboration, I had formed an umbilical attachment to Tom, a man who had given me so much—his time, his expertise, his support—and asked nothing return. There was an unspoken understanding that I was part of the mission, *his* mission, to defend the defenseless. No one would ever live up to his preposterous standards, but I spent every day imagining how I might do a little better, and the dalba study showed that I could run a trial on my own. He'd handed the baton to me, and I didn't drop it.

There were still gaps in my understanding of drug development, and I dreaded the day Tom wouldn't be around to help patients and their doctors. But as the years passed, I became more confident that he would never retire. He couldn't. Medicine energizes him. It gives him life. Without it, he would be lost. And for that, we are all fortunate. Tom will never retreat from the mission because superbugs will never go away. In fact, we can expect to see far more of them in the years to come, but they're up against a formidable opponent, one who is quietly devising a wondrous new plan of attack.

Tom leaned forward in his chair, tapped his computer, and pulled up YouTube. "Do you mind?" he asked, knowing I wouldn't. Vivaldi, Beethoven, it was all the same to me. Tom found a song, pressed Play, and returned to the yellow paper before him. There was a bemused look on his face, one that I hadn't seen in a long time. "Let's get started," he said, staring at the drawing. Beneath the words *New Protocol*, I jotted down a half dozen genes that we could target with CRISPR, the molecular scalpel. Tom pulled a pen from his breast pocket and studied my list as the room filled gently with the sounds of a reluctant vocalist encouraging us to take it to the limit, one more time.

Acknowledgments

SUPERBUGS WAS TRULY a team effort. I am fortunate to work with two outstanding editors, Megan Newman and Nina Shield, a thoughtful and supportive agent, Scott Waxman (has it really been a decade?), as well as a talented group at Penguin Random House, including Casey Maloney, Lindsay Gordon, Farin Schlussel, Hannah Steigmeyer, and Allyssa Kasoff. I also want to thank my teams in New Smyrna Beach, Irvington, Mahopac, Arlington, Weill Cornell, and Allergan for their patience and encouragement, and I want to high-five a copyeditor who knows more about the Eagles than anyone should. *Take it easy!*

I am lucky to have a group of close friends—Rachel, Charlie, John, and Ben—who provide a daily sounding board for my hot takes, and I am indebted to Kevin Doughten for his much-needed humor and to Eden Laase for her thorough fact-checking. Ultimately this is a story about patients, and I am most grateful to the men and women who took a chance on our clinical trial. Thank you for your generosity and courage.

Heather, I'm not sure I'll ever fully express what you mean to me, but every day with you is a good one. Thank you for showing me what it means to be tough and kind. I can't wait to see what happens next.

A Note on Sources

SUPERBUGS WAS CREATED over the past five years from more than six hundred primary and secondary sources as well as dozens of on- and off-the-record interviews. I am indebted to the team at the Weill Cornell Medicine library for their time and expertise in tracking down many of the relevant documents, which also included newspaper clippings, magazine articles, and blog posts. A complete bibliography would be another book in itself. Below, I have selected references that were cited or that contributed directly to the manuscript.

Notes

Prologue

2 **outrageously toxic**: A. Ordooei Javan, S. Shokouhi, and Z. Sahraei, "A Review on Colistin Nephrotoxicity," *European Journal of Clinical Pharmacology* 71, no. 7 (2015): 801–10.

2 **so effective**: C. Nathan and O. Cars, "Antibiotic Resistance—Problems, Progress, and Prospects," *New England Journal of Medicine* 371, no. 19 (2014): 1761–63.

2 **twenty thousand people die every year**: Centers for Disease Control and Prevention, *Antibiotic Resistance Threats in the United States, 2013*, April 23, 2013, www.cdc.gov/drugresistance/pdf/ar-threats-2013-508.pdf.

3 **special precautions**: S. A. Clock et al., "Contact Precautions for Multidrug-Resistant Organisms: Current Recommendations and Actual Practice," *American Journal of Infection Control* 38, no. 2 (2010): 105–11.

4 **persistent problem**: C. L. Ventola, "The Antibiotic Resistance Crisis: Part 1: Causes and Threats," *Pharmacy and Therapeutics (P&T)* 40, no. 4 (2015): 277–83.

4 **prescribing practices**: A. Nicholson et al., "The Knowledge, Attitudes and Practices of Doctors Regarding Antibiotic Resistance at a Tertiary Care Institution in the Caribbean," *Antimicrobial Resistance and Infection Control* 7 (2018): 23.

4 **indiscriminate use**: T. F. Landers et al., "A Review of Antibiotic Use in Food Animals: Perspective, Policy, and Potential," *Public Health Reports* 127, no. 1 (2012): 4–22.

4 **everywhere**: H. S. Gold and R. C. Moellering, "Antimicrobial-Drug Resistance," *New England Journal of Medicine* 335, no. 19 (1996): 1445–53.

4 **treatable just a few years ago**: V. Idemyor, "Antimicrobial Drug Resistance Among Common Pathogens in American Hospitals: When Will the Microbe Stop Winning?," *Journal of the National Medical Association* 90, no. 1 (1998): 10–12.

4 **Nobel laureate**: P. Zhao, "The 2009 Nobel Prize in Chemistry: Thomas A. Steitz and the Structure of the Ribosome," *Yale Journal of Biology and Medicine* 84, no. 2 (2011): 125–29.

5   **infectious disease doctor**: P. Farmer et al., "Community-Based Approaches to HIV Treatment in Resource-Poor Settings," *Lancet* 358, no. 9279 (2001): 404–9.

5   **failed studies**: R. L. Engler, "Misrepresentation and Responsibility in Medical Research," *New England Journal of Medicine* 317, no. 22 (1987): 1383–99.

5   **ethical lapses**: H. T. Shapiro and E. M. Meslin, "Ethical Issues in the Design and Conduct of Clinical Trials in Developing Countries," *New England Journal of Medicine* 345, no. 2 (2001): 139–42.

CHAPTER 1: The Fog of War

9   **Royal Army Medical Corps**: R. Hare, "The Scientific Activities of Alexander Fleming, Other Than the Discovery of Penicillin," *Medical History* 27, no. 4 (1983): 347–72.

9   **terrible fates:** S. Sabbatani and S. Fiorino, "The Treatment of Wounds During World War I," *Le Infezioni in Medicina* 25, no. 2 (2017): 184–92.

9   **tetanus**: P. C. Wever and L. van Bergen, "Prevention of Tetanus During the First World War," *Medical Humanities* 38, no. 2 (2012): 78–82.

9   **wound-research laboratory**: E. Lax, *The Mold in Dr. Florey's Coat: The Story of the Penicillin Miracle* (New York: Henry Holt, 2004).

9   **losses were heavy**: E. Jones, "Terror Weapons: The British Experience of Gas and Its Treatment in the First World War," *War in History* 21, no. 3 (2014): 355–75.

9   **August**: B. Tuchman, *The Guns of August* (New York: Random House, 2009).

9   **retreat**: T. Zuber, *The Battle of the Frontiers: Ardennes 1914* (Stroud, UK: History Press, 2010).

10   **British Expeditionary Force**: A. Rawson, *British Expeditionary Force: The 1914 Campaign* (Barnsley, UK: Pen and Sword Military, 2014).

10   **bloodiest engagements**: R. A. Pollock, "Triage and Management of the Injured in World War I: The Diuturnity of Antoine De Page and a Belgian Colleague," *Craniomaxillofacial Trauma & Reconstruction* 1, no. 1 (2008): 63–70.

10   **abandon plans**: G. Scharf, "Holding the Torch up High," *South African Journal of Surgery* 55, no. 3 (2017): 67.

10   **extraordinary cost**: H. Herwig, *The Marne, 1914: The Opening of World War I and the Battle That Changed the World* (New York: Random House, 2009).

10   **shreds of uniform**: G. Macfarlane, *Alexander Fleming: The Man and the Myth* (New York: Oxford University Press, 1985).

10   **swatch of fabric**: A. Fleming, "Antiseptics, Old and New," *Proceedings of the Staff Meetings of the Mayo Clinic* 21 (1946): 67–75.

10   **so many**: G. D. Shanks, "How World War I Changed Global Attitudes to War and Infectious Diseases," *Lancet* 384, no. 9955 (2014): 1699–707.

10   **tetanus**: J. Boyd, "Tetanus in Two World Wars," *Proceedings of the Royal Society of Medicine* 52, no. 2 (1959): 109–10.

10   **embedded in British military uniforms**: K. Brown, *Penicillin Man: Alexander Fleming and the Antibiotic Revolution* (Stroud, UK: History Press, 2005).

10   **his laboratory**: Lax, *Mold in Dr. Florey's Coat.*

10   **injured soldiers**: Herwig, *Marne, 1914.*

10   **Fleming's ingenuity**: Macfarlane, *Alexander Fleming.*

10   **tubes**: A. Fleming, "A British Medical Association Lecture on Vaccine Therapy in Regard to General Medical Practice," *British Medical Journal* 1, no. 3138 (1921): 255–59.

11   **antiseptic fluid**: Lax, *The Mold in Dr. Florey's Coat*.

11   **discharge from the leg**: J. G. Adami et al., "Combined Inquiry into the Presence of Diphtheria and Diphtheroid Bacilli in Open Wounds," *Canadian Medical Association Journal* 8, no. 9 (1918): 769–85.

11   **soup of pus**: A. Fleming, "The Bactericidal Power of Human Blood and Some Methods of Altering It," *Proceedings of the Royal Society of Medicine* 21, no. 5 (1928): 859–68.

11   **lockjaw**: Boyd, "Tetanus," 109–10.

11   **tetanus**: K. C. Dittrich and B. Keilany, "Tetanus: Lest We Forget," *Canadian Journal of Emergency Medicine* 3, no. 1 (2001): 47–50.

11   **risus sardonicus**: S. Levin, "Risus Sardonicus," *Adler Museum Bulletin* 16, no. 1 (1990): 19–22.

11   **spores**: Shanks, "How World War I Changed Global Attitudes," 1699–707.

11   **anaerobic**: P. Finkelstein et al., "Tetanus: A Potential Public Health Threat in Times of Disaster," *Prehospital and Disaster Medicine* 32, no. 3 (2017): 339–42.

11   **Even brief exposure**: V. Fredette, C. Planté, and A. Roy, "Numerical Data Concerning the Sensitivity of Anaerobic Bacteria to Oxygen," *Journal of Bacteriology* 94, no. 6 (1967): 2012–17.

11   **flourishing**: Shanks, "How World War I Changed Global Attitudes," 1699–707.

11   **recesses of the wound**: Lax, *The Mold in Dr. Florey's Coat*.

11   **Almroth Wright**: J. R. Matthews, "Almroth Wright, Vaccine Therapy, and British Biometrics: Disciplinary Expertise Versus Statistical Objectivity," *Clio Medica* 67 (2002): 125–47.

11   **French widow**: Brown, *Penicillin Man*.

11   **hormonal gland disorder**: Macfarlane, *Alexander Fleming*.

12   **impassioned plea**: W. Gillespie, "Paul Ehrlich and Almroth Wright," *West of England Medical Journal* 106, no. 4 (1991): 107, 18.

12   **Britain had declared war**: E. W. Meynell, "Some Account of the British Military Hospitals of World War I at Etaples, in the Orbit of Sir Almroth Wright," *Journal of the Royal Army Medical Corps* 142, no. 1 (1996): 43–47.

12   **unpopular with many doctors**: Brown, *Penicillin Man*.

12   **unconvinced**: J. Murray, "Sir Alfred Keogh: Doctor and General." *Irish Medical Journal* 80, no. 12 (1987): 427–32.

12   **more research was needed**: J. S. Blair, "Sir Alfred Keogh—The Great War," *Journal of the Royal Army Medical Corps* 154, no. 4 (2008): 273–74.

12   **now found himself**: R. L. Atenstaedt, "The Organisation of the RAMC During the Great War," *Journal of the Royal Army Medical Corps* 152, no. 2 (2006): 81–85.

12   **consumed**: Fleming, "A British Medical Association Lecture on Vaccine Therapy," 255–59.

12   **hard to kill**: A. Fleming, "The Bactericidal Power of Human Blood and Some Methods of Altering It," *Proceedings of the Royal Society of Medicine* 21, no. 5 (1928): 859–68.

12   **For some of his infected men**: W. C. Hanigan, "Neurological Surgery During the Great War: The Influence of Colonel Cushing," *Neurosurgery* 23, no. 3 (1988): 283–94.

12   **antiseptic school**: W. Spink, *Infectious Diseases: Prevention and Treatment in the Nineteenth and Twentieth Centuries* (Minneapolis: University of Minnesota Press, 1979).

12   **antiseptics worked well in theory**: Fleming, "Antiseptics, Old and New," 67–75.

12   **hunch**: Hare, "Scientific Activities of Alexander Fleming," 347–72.

12   **His theory**: P. M. Mazumdar, "Fleming as Bacteriologist: Alexander Fleming," *Science* 225, no. 4667 (1984): 1140.

13   **Something about the periphery**: A. R. Munroe, A. G. Fleming, and R. M. Janes, "Wound Flora in Relation to Secondary Suture," *British Medical Journal* 1, no. 2980 (1918): 173.

13   **life before medicine**: Lax, *The Mold in Dr. Florey's Coat*.

13   **glass-blowing**: Ibid.

13   **scampering mice**: Macfarlane, *Alexander Fleming*.

13   **dreaming**: Y. Manzano and J. Manzano, "Homage to Sir Alexander Fleming," *Hispalis Medica* 12, no. (1955): 237–39.

13   **still in its infancy**: F. Marti-Ibanez, "The Meaning of Greatness: Sir Alexander Fleming—In Memoriam," *Antibiotics and Chemotherapy* 5, no. 4 (1955): 177–81.

13   **Antiseptics**: Fleming, "Antiseptics, Old and New," 67–75.

13   **not simply useless**: V. D. Allison, "Personal Recollections of Sir Almroth Wright and Sir Alexander Fleming," *Ulster Medical Journal* 43, no. 2 (1974): 89–98.

13   **trying to speak with him**: G. Thompson, ed., *Nobel Prizes That Changed Medicine* (London: Imperial College Press, 2012).

13   **Fleming returned to London**: A. Geddes, "80th Anniversary of the Discovery of Penicillin: An Appreciation of Sir Alexander Fleming," *International Journal of Antimicrobial Agents* 32, no. 5 (2008): 373.

13   **known in academic circles**: A. Fleming, "Lysozyme: President's Address," *Proceedings of the Royal Society of Medicine* 26, no. 2 (1932): 71–84.

14   **just up Praed Street**: Brown, *Penicillin Man*.

14   **chance observation**: Macfarlane, *Alexander Fleming*.

14   **killed in the presence of a fungus**: Brown, *Penicillin Man*.

14   **discarded petri dish**: M. Lobanovska and G. Pilla, "Penicillin's Discovery and Antibiotic Resistance: Lessons for the Future?," *Yale Journal of Biology and Medicine* 90, no. 1 (2017): 135–45.

14   **dubbed it penicillin**: A. Fleming, "The Development and Use of Penicillin," *Chicago Medical School Quarterly* 7, no. 2 (1946): 20–28.

14   **he sent his findings**: A. Fleming, "Classics in Infectious Diseases: On the Antibacterial Action of Cultures of a Penicillium, with Special Reference to Their Use in the Isolation of *B. influenzae* by Alexander Fleming," reprinted from the *British Journal of Experimental Pathology* 10 (1929): 226–36; *Reviews of Infectious Diseases* 2, no. 1 (1980): 129–39.

14   **not yet clear**: A. Fleming, "The Assay of Penicillin in the Days Before It Was Concentrated," *Bulletin of the Health Organisation, League of Nations* 12, no. 2 (1945): 250–52.

14   **resigned to the fact**: Brown, *Penicillin Man*.

14   **a valuable tool**: A. Fleming and W. Walters, "Penicillin in Surgery," *Lancet* 2, no. 6474 (1947): 479.

15   **was not written in a way**: Fleming, "Classics in Infectious Diseases: On the Antibacterial Action of Cultures of a Penicillium," 129–39.

15   **his chemical reagents**: Brown, *Penicillin Man*.

15   **misidentified his fungus**: Lax, *The Mold in Dr. Florey's Coat*.

15   **responsible for the influenza pandemic**: E. Tognotti, "Scientific Triumphalism and Learning from Facts: Bacteriology and the 'Spanish Flu' Challenge of 1918," *Social History of Medicine* 16, no. 1 (2003): 97–110.

15 **cases of influenza**: R. H. Ivy, "The Influenza Epidemic of 1918: Personal Experience of a Medical Officer in World War I," *Military Medicine* 125 (1960): 620–22.

15 **urgent need**: K. R. Robinson, "The Role of Nursing in the Influenza Epidemic of 1918–1919," *Nursing Forum* 25, no. 2 (1990): 19–26.

15–16 **stumbled upon a rare strain of fungus**: Fleming, "Development and Use of Penicillin," 20–28.

16 **abandoned work**: Brown, *Penicillin Man*.

16 **first mass-produced**: W. Rosen, *Miracle Cure: The Creation of Antibiotics and the Birth of Modern Medicine* (New York: Viking, 2017).

CHAPTER 2: A Golden Era

17 **consuming antibiotics**: Lax, *The Mold in Dr. Florey's Coat*.

17 **Sudanese mummies**: E. J. Bassett et al., "Tetracycline-Labeled Human Bone from Ancient Sudanese Nubia (A.D. 350)," *Science* 209, no. 4464 (1980): 1532–34.

17 **Dakhla Oasis**: M. Cook, E. Molto, and C. Anderson, "Fluorochrome Labelling in Roman Period Skeletons from Dakhleh Oasis, Egypt," *American Journal of Physical Anthropology* 80, no. 2 (1989): 137–43.

17 **an array of antibiotics**: R. I. Aminov, "A Brief History of the Antibiotic Era: Lessons Learned and Challenges for the Future," *Frontiers in Microbiology* 1 (2010): 134.

17 **red soils in Jordan**: J. O. Falkinham et al., "Proliferation of Antibiotic-Producing Bacteria and Concomitant Antibiotic Production as the Basis for the Antibiotic Activity of Jordan's Red Soils," *Applied and Environmental Microbiology* 75, no. 9 (2009): 2735–41.

18 **bleach**: T. M. Barnes and K. A. Greive, "Use of Bleach Baths for the Treatment of Infected Atopic Eczema," *Australasian Journal of Dermatology* 54, no. 4 (2013): 251–58.

18 **without killing us**: B. N. Tse et al., "Challenges and Opportunities of Nontraditional Approaches to Treating Bacterial Infections," *Clinical Infectious Diseases* 65, no. 3 (2017): 495–500.

18 **scientists often quibble**: N. Wald-Dickler, P. Holtom, and B. Spellberg, "Busting the Myth of 'Static vs. Cidal': A Systemic Literature Review," *Clinical Infectious Diseases* 66, no. 9 (2018): 1470–74.

18 **bacteria and viruses were different**: G. G. Vinson, "Possible Chemical Nature of Tobacco Mosaic Virus," *Science* 79, no. 2059 (1934): 548–49.

18 **new antibiotics**: Rosen, *Miracle Cure*.

18 **golden era**: M. Lobanovska and G. Pilla, "Penicillin's Discovery and Antibiotic Resistance: Lessons for the Future?," *Yale Journal of Biology and Medicine* 90, no. 1 (2017): 135–45.

18 **half of the drugs in use today**: J. Davies, "Where Have All the Antibiotics Gone?," *Canadian Journal of Infectious Diseases and Medical Microbiology* 17, no. 5 (2006): 287–90.

18 **just eight people**: Rosen, *Miracle Cure*.

18 **one hundred**: Ibid.

18 **broad-spectrum antibiotics**: Ibid.

18–19 **long-standing policy**: D. Rapoport, "Physicians and the Pharmaceutical Industry: Under the Influence?," *Canadian Medical Association Journal* 152, no. 1 (1995): 15.

19   **they could make more money**: Rosen, *Miracle Cure*.

19   **Jawetz took stock of the progress**: E. Jawetz, "Antimicrobial Chemotherapy," *Annual Review of Microbiology* 10 (1956): 85–114.

19   **pointed out a few problems**: M. Finland and L. Weinstein, "Complications Induced by Antimicrobial Agents," *New England Journal of Medicine* 248, no. 6 (1953): 220–26.

19   **urged caution**: Ibid.

19   **linked to aplastic anemia**: A. W. Johnston, "Aplastic Anaemia Following Treatment with Chloramphenicol: Transfusion of Polycythaemic Blood Using Sequestrene," *Lancet* 267, no. 6833 (1954): 319.

19   **toxic side effects**: W. Dameshek, "Chloramphenicol—A New Warning," *Journal of the American Medical Association* 174 (1960): 1853–54.

19   **built enzymes**: E. P. Abraham and E. Chain, "An Enzyme from Bacteria Able to Destroy Penicillin—1940," *Reviews of Infectious Diseases* 10, no. 4 (1988): 677–78.

20   **Burnet**: M. Burnet, *Natural History of Infectious Disease* (Cambridge: Cambridge University Press, 1962).

20   **No new classes**: A. R. Coates, G. Halls, Y. Hu, "Novel Classes of Antibiotics or More of the Same?," *British Journal of Pharmacology* 163, no. 1 (2011): 184–94.

20   **a breakthrough**: R. D. Fleischmann et al., "Whole-Genome Random Sequencing and Assembly of *Haemophilus influenzae* Rd.," *Science* 269, no. 5223 (1995): 496–512.

20   **redefine drug development**: J. C. Venter et al., "Shotgun Sequencing of the Human Genome," *Science* 280, no. 5369 (1998): 1540–42.

20   **genomic bandwagon**: D. J. Payne et al., "Drugs for Bad Bugs: Confronting the Challenges of Antibacterial Discovery," *Nature Reviews Drug Discovery* 6, no. 1 (2007): 29–40.

21   **none was useful in humans**: A. Miller and P. Miller, eds., *Emerging Trends in Antibacterial Discovery: Answering the Call to Arms* (Norfolk, UK: Caister Academic Press, 2011).

21   **patents filed**: B. N. Tse, "Challenges and Opportunities," 495–500.

22   **not very profitable**: B. Spellberg, J. G. Bartlett, and D. N. Gilbert. "The Future of Antibiotics and Resistance," *New England Journal of Medicine*" 368, no. 4 (2013): 299–302.

22   **For antibacterial drugs**: B. Spellberg, "The Future of Antibiotics," *Critical Care* 18, no. 3 (2014): 228.

23   **How far should we go**: M. Rosenblatt, "The Large Pharmaceutical Company Perspective," *New England Journal of Medicine* 376, no. 1 (2017): 52–60.

23   **rational drug design**: Q. Q. Ma et al., "Rational Design of Cationic Antimicrobial Peptides by the Tandem of Leucine-Rich Repeat," *Amino Acids* 44, no. 4 (2013): 1215–24.

23   **see what happens**: M. S. Chegkazi et al., "Rational Drug Design Using Integrative Structural Biology," *Methods in Molecular Biology* 1824 (2018): 89–111.

23   **innovative programmers and scientists**: W. Duch, K. Swaminathan, and J. Meller, "Artificial Intelligence Approaches for Rational Drug Design and Discovery," *Current Pharmaceutical Design* 13, no. 14: 1497–508.

24   **Dalba was made**: R. Hermann et al., "Synthesis and Antibacterial Activity of Derivatives of the Glycopeptide Antibiotic A-40926 and Its Aglycone," *Journal of Antibiotics* 49, no. 12 (1996): 1236–48.

24   **careful not to tamper with**: M. Steiert and F. J. Schmitz, "Dalbavancin (Biosearch Italia/Versicor)," *Current Opinion in Investigational Drugs* 3, no. 2 (2002): 229–33.

25   **predicted a 2005 launch**: Ibid.

25   **the drug was expected to go viral**: Ibid.

25   **trying to show**: S. Ramdeen and H. W. Boucher, "Dalbavancin for the Treatment of Acute Bacterial Skin and Skin Structure Infections," *Expert Opinion on Pharmacotherapy* 16, no. 13 (2015): 2073–81.

25   **threw in the towel**: Ibid.

25   **eventually purchased by Allergan**: B. Das et al., "Review: Dalbavancin—A Novel Lipoglycopeptide Antimicrobial for Gram Positive Pathogens," *Pakistan Journal of Pharmacolological Sciences* 21, no. 1 (2008): 78–87.

25   **finally approved**: T. M. Khadem, R. P. van Manen, and J. Brown, "How Safe Are Recently FDA-Approved Antimicrobials? A Review of the FDA Adverse Event Reporting System Database," *Pharmacotherapy* 34, no. 12 (2014): 1324–29.

25   **molecular machinery**: M. A. Pfaller, "Antifungal Drug Resistance: Mechanisms, Epidemiology, and Consequences for Treatment," *American Journal of Medicine* 125, no. 1 supp. (2012): S3–13.

CHAPTER 3: The Lucky Grenadier

29   **Domagk wrote from a coastal town**: E. Grundmann, *Gerhard Domagk: The First Man to Triumph over Infectious Diseases* (Münster, Germany.: LIT Verlag, 2005).

29   **closer to danger**: Ibid.

29   **swept up in a wave**: Ibid.

29   **specializing in grenades**: Ibid.

29   **traded his white coat**: T. N. Raju, "The Nobel Chronicles—1939: Gerhard Domagk (1895–1964)," *Lancet* 353, no. 9153 (1999): 681.

30   **First Battle of Ypres**: J. Sheldon, *The German Army at Ypres—1914* (Barnsley, UK: Pen and Sword Military, 2011).

30   **Domagk survived**: Grundmann, *Gerhard Domagk*.

30   **a barn had been converted**: T. Hager, *The Demon Under the Microscope: From Battlefield Hospitals to Nazi Labs, One Doctor's Heroic Search for the World's First Miracle Drug* (New York: Broadway Books, 2007).

30   **transported the dead and dying**: Ibid.

30   **talented medical student**: Ibid.

31   **met his future mentor**: Grundmann, *Gerhard Domagk*.

31   **high-minded thinker**: M. A. Kinch, *Prescription for Change: The Looming Crisis in Drug Development* (Chapel Hill: University of North Carolina Press, 2016).

31   **"Whatever contributes"**: Grundmann, *Gerhard Domagk*.

31   **he could spot four-leaf clovers**: Hager, *The Demon Under the Microscope*.

31   **The family motto**: Ibid.

31   **took a job**: J. Lesch, *The First Miracle Drugs: How the Sulfa Drugs Transformed Medicine* (New York: Oxford University Press, 2007).

31   **academia had close ties**: Hager, *The Demon Under the Microscope*.

32   **Domagk's boss**: Ibid.

32   **two-year renewable contract**: Raju, "Nobel Chronicles—1939," 681.

32   **chemical screens**: I. Sherman, *Drugs That Changed the World: How Therapeutic Agents Shaped Our Lives* (Boca Raton, FL: CRC Press, 2016).

32   **spent most of his days**: Hager, *The Demon Under the Microscope.*

32   **wool dyes**: Grundmann, *Gerhard Domagk.*

33   **Kl-695**: Hager, *The Demon Under the Microscope.*

33   **a cheap and practical dye**: G. Domagk, "Twenty-five Years of Sulfonamide Therapy," *Annals of the New York Academy of Sciences* 69, no. 3 (1957): 380–84.

33   **The results were striking**: Hager, *The Demon Under The Microscope.*

33   **Kl-730**: Grundmann, *Gerhard Domagk.*

33   **couldn't patent every variant**: "Obituary Notices: G. Domagk," *British Medical Journal* 1, no. 5391 (1964): 1189.

33   **filed a patent**: Hager, *The Demon Under the Microscope.*

34   **gas was used**: B. Carruthers, *Poland 1939: The Blitzkrieg Unleashed* (Slough, UK: Archive Media Publishing, 2011).

34   **professional group**: R. J. Lifton, *The Nazi Doctors: Medical Killing and the Psychology of Genocide* (New York: Basic Books, 1988).

34   **revelations began**: V. Spitz, *Doctors from Hell: The Horrific Account of Nazi Experiments on Humans* (Boulder, CO: Sentient Publications, 2005).

34   **military tribunals**: Ibid.

34   **grotesque role**: Lifton, *Nazi Doctors.*

34   **the twenty-three defendants**: Spitz, *Doctors from Hell.*

34   **scientists into assassins**: Lifton, *Nazi Doctors.*

35   **"experimental persons"**: Spitz, *Doctors from Hell.*

35   **combining sulfanilamide with other molecules**: Domagk, "Twenty-five Years of Sulfonamide Therapy."

36   **weren't convinced**: Hager, *The Demon Under the Microscope.*

36   **"Jewish science"**: P. Ball, *Serving the Reich: The Struggle for the Soul of Physics Under Hitler* (Chicago: University of Chicago Press, 2014).

36   **forbade him from accepting**: Raju, "Nobel Chronicles—1939," 681.

36   **were experimented on**: Spitz, *Doctors from Hell.*

36   **procedure was repeated**: Ibid.

36   **In a third group**: Ibid.

37   **hopped or limped**: P. Weindling et al., "The Victims of Unethical Human Experiments and Coerced Research Under National Socialism," *Endeavour* 40, no. 1 (2016): 1–6.

37   **survivors testified**: Spitz, *Doctors from Hell.*

37   **soldiers with actual battlefield infections**: Ibid.

37   **sentenced to death**: Lifton, *Nazi Doctors.*

37   **physiologist and ethicist**: A. Gaw, "Reality and Revisionism: New Evidence for Andrew C. Ivy's Claim to Authorship of the Nuremberg Code," *Journal of the Royal Society of Medicine* 107, no. 4 (2014): 138–43.

37   **where he studied**: L. A. Temme, "Ethics in Human Experimentation: The Two Military Physicians Who Helped Develop the Nuremberg Code," *Aviation, Space, and Environmental Medicine* 74, no. 12 (2003): 1297–300.

37   **ten-point framework**: E. Shuster, "Fifty Years Later: The Significance of the Nuremberg Code," *New England Journal of Medicine* 337, no. 20 (1997): 1436–40.

38   **appropriate supervision**: E. Shuster, "American Doctors at the Nuremberg Medical Trial," *American Journal of Public Health* 108, no. 1 (2018): 47–52.

38   **"Not to my knowledge"**: Spitz, *Doctors from Hell.*

CHAPTER 4: Embedded

39 **tapped to move**: A. M. Brandt, "Racism and Research: The Case of the Tuskegee Syphilis Study," *Hastings Center Report* 8, no. 6 (1978): 21–29.

39 **recruiting black men**: J. Park, "Historical Origins of the Tuskegee Experiment: The Dilemma of Public Health in the United States," *Korean Journal of Medical History* 26, no. 3 (2017): 545–78.

39 **cheap and simple project**: J. Jones, *Bad Blood: The Tuskegee Syphilis Experiment* (New York: Free Press, 1993).

39 **"a night with Venus and a lifetime with Mercury"**: A. D. Opina and A. A. Tafur, "A Night with Venus, a Lifetime with Mercury: A Case of Multiple Intracranial Aneurysms," *American Journal of the Medical Sciences* 343, no. 6 (2012): 498–500.

39 **ongoing work**: Brandt, "Racism and Research," 21–29.

40 **one of the few in the country with subspecialty training**: F. Gray, *The Tuskegee Syphilis Study* (Montgomery, AL: NewSouth Books, 2002).

40 **Eighty-two percent were black**: Ibid.

40 **the illiteracy rate**: Jones, *Bad Blood*.

40 **malnutrition**: Ibid.

40 **arrived in Alabama**: Ibid.

40 **to announce that government doctors had arrived**: D. Hardy, *I'm from the Government and I'm Here to Kill You: The True Human Cost of Official Negligence* (New York: Skyhorse, 2017).

40 **in conjunction with local officials**: Park, "Historical Origins of the Tuskegee Experiment," 545–78.

41 **samples that were shipped**: Jones, *Bad Blood*.

41 **notified by mail**: Ibid.

41 **"bad blood"**: Ibid.

41 *bad blood* **actually referred to**: Ibid.

41 **"details of the puncture"**: Hardy, *I'm from the Government*.

41 **he wrote to his patients**: Jones, *Bad Blood*.

41 **"flair for framing letters"**: Ibid.

41 **The ruse worked**: Gray, *Tuskegee Syphilis Study*.

42 **terminating the study**: Jones, *Bad Blood*.

42 **promoted**: E. Emanuel and C. Grady, eds., *The Oxford Textbook of Clinical Research Ethics* (New York: Oxford University Press, 2011).

42 **no end date**: Jones, *Bad Blood*.

42 **treatment was mandated**: Ibid.

42 **treating syphilis**: Brown, *Penicillin Man*.

43 **consented to autopsies**: Jones, *Bad Blood*.

43 **outrage was swift**: W. J. Curran, "The Tuskegee Syphilis Study," *New England Journal of Medicine* 289, no. 14 (1973): 730–31.

43 **"controlled genocide"**: Jones, *Bad Blood*.

43 **openly discussed**: Brandt, "Racism and Research," 21–29.

43 **doctors reported**: J. J. Peters et al., "Untreated Syphilis in the Male Negro; Pathologic Findings in Syphilitic and Nonsyphilitic patients," *Journal of Chronic Diseases* 1, no. 2 (1955): 127–48.

43 **In return for**: Jones, *Bad Blood*.

43 **blame was scattered**: Curran, "The Tuskegee Syphilis Study," 730–31.

43–44 **simply never offered**: R. M. White, "Unraveling the Tuskegee Study of Untreated Syphilis," *Archives of Internal Medicine* 160, no. 5 (2000): 585–98.

44 **the experiment should continue**: Brandt, "Racism and Research," 21–29.

44 **state-sponsored abuse**: E. Y. Adashi, L. B. Walters, and J. A. Menikoff, "The Belmont Report at 40: Reckoning with Time," *American Journal of Public Health* 108, no. 10 (2018): 1345–48.

44 **drawn up and allowed to continue**: Curran, "The Tuskegee Syphilis Study," 730–31.

44 **no protocol**: Brandt, "Racism and Research," 21–29.

CHAPTER 5: Safeguards

45 **stays in your body**: E. Seltzer et al., "Once-Weekly Dalbavancin Versus Standard-of-Care Antimicrobial Regimens for Treatment of Skin and Soft-Tissue Infections," *Clinical Infectious Diseases* 37, no. 10 (2003): 1298–303.

46 **the trial should continue**: Brandt, "Racism and Research," 21–29.

47 **Beecher published a paper**: H. K. Beecher, "Ethics and Clinical Research," *New England Journal of Medicine* 274, no. 24 (1966): 1354–60.

47 **"obedience"**: S. Gibson, "Obedience Without Orders: Expanding Social Psychology's Conception of 'Obedience,'" *British Journal of Social Psychology* (e-pub, 2018): http://doi:10.1111/bjso.12272.

47 **famously tricking**: J. Laurent, "Milgram's Shocking Experiments: A Case in the Social Construction of 'Science,'" *Indian Journal of History of Science* 22, no. 3 (1987): 247–72.

47 **cancer cells were injected**: Beecher, "Ethics and Clinical Research," 1354–60.

47 **administration of a virus**: Ibid.

47 **unethical, outrageous studies**: D. J. Rothman, "Ethics and Human Experimentation: Henry Beecher Revisited," *New England Journal of Medicine* 317, no. 19 (1987): 1195–97.

47 **he had served in military field hospitals**: F. Benedetti, "Beecher as Clinical Investigator: Pain and the Placebo Effect," *Perspectives in Biology and Medicine* 59, no. 1 (2016): 3745.

47 **calm injured soldiers**: D. D. Price, D. G. Finniss, and F. Benedetti, "A Comprehensive Review of the Placebo Effect: Recent Advances and Current Thought," *Annual Review of Psychology* 59 (2008): 565–90.

47 **power of the placebo effect**: H. K. Beecher, "The Powerful Placebo," *Journal of the American Medical Association* 159, no. 17 (1955): 1602–6.

47 **continued to study**: H. K. Beecher et al., "The Effectiveness of Oral Analgesics (Morphine, Codeine, Acetylsalicylic Acid) and the Problem of Placebo 'Reactors' and 'Non-Reactors,'" *Journal of Pharmacology and Experimental Therapeutics* 109, no. 4 (1953): 393–400.

47 **he is best known**: D. S. Jones, C. Grady, and S. E. Lederer, "'Ethics and Clinical Research'—The 50th Anniversary of Beecher's Bombshell," *New England Journal of Medicine* 374, no. 24 (2016): 2393–98.

47 **Beecher wrote in his conclusion**: Beecher, "Ethics and Clinical Research," 1354–60.

47 **He called on**: Ibid.

48 **the focal point**: Jones, Grady, and Lederer, "The 50th Anniversary of Beecher's Bombshell," 2393–98.

48   **review committees**: L. Stark, *Behind Closed Doors: IRBs and the Making of Ethical Research* (Chicago: University of Chicago Press, 2012).

48   **the modern institutional review board**: R. Novak, "Review of Human Subjects Research: Options Available to the Institutional Review Board," *Grants* 6, no. 4 (1983): 225–31.

48   **experts could evaluate**: S. E. Lind, "The Institutional Review Board: An Evolving Ethics Committee," *Journal of Clinical Ethics* 3, no. 4 (1992): 278–82.

48   **National Research Act**: SENATE USC, (Reprint of) National Research Act, Conference Report, *Drug Research Reports* 17, no. 28 (1974): 51–536.

48   **diverse group**: L. Stark, *Behind Closed Doors*.

49   **behind closed doors**: Ibid.

CHAPTER 6: Variables

51   **convene without researchers**: L. Stark, *Behind Closed Doors*.

51   **payment system that altered**: K. Quinn, "After the Revolution: DRGs at Age 30," *Annals of Internal Medicine* 160, no. 6 (2014): 426–29.

51   **length of stay for a patient dropped**: M. L. Barnett et al., "Home-to-Home Time—Measuring What Matters to Patients and Payers," *New England Journal of Medicine* 377, no. 1 (2017): 4–6.

52   **must include**: L. Stark, *Behind Closed Doors*.

CHAPTER 7: Deferment

55   **The drug was discovered to cause**: M. D. Blum, D. J. Graham, and C. A. McCloskey, "Temafloxacin Syndrome: Review of 95 Cases," *Clinical Infectious Diseases* 18, no. 6 (1994): 946–50.

57   **Bernadine Healy**: D. E. Shalala, "Retrospective: Bernadine Healy (1944–2011)," *Science* 333, no. 6051 (2011): 1836.

58   **first female director**: J. Palca, "Bernadine Healy: A New Leadership Style at NIH," *Science* 253, no. 5024 (1991): 1087–89.

CHAPTER 8: Oversight

60   **this paradigm is changing**: T. Guillard et al., "Antibiotic Resistance and Virulence: Understanding the Link and Its Consequences for Prophylaxis and Therapy," *Bioessays* 38, no. 7 (2016): 682–93.

60   **efflux pumps**: S. Jang, "Multidrug Efflux Pumps in *Staphylococcus aureus* and Their Clinical Implications," *Journal of Microbiology* 54, no. 1 (2016): 1–8.

61   **Spellberg and his team have noted**: Spellberg, Bartlett, and Gilbert, "The Future of Antibiotics and Resistance," 299–302.

62   **handful of experts**: P. Hilts, *Protecting America's Health: The FDA, Business and One Hundred Years of Regulation* (Chapel Hill: University of North Carolina Press, 2004).

63   **signed into law**: Ibid.

64   **make rotten vegetables appear ripe**: Ibid.

64   **added to milk**: D. Blum, *The Poison Squad: One Chemist's Single-Minded Crusade for Food Safety at the Turn of the Twentieth Century* (New York: Penguin Press, 2018).

64   **did not have the power**: Hilts, *Protecting America's Health.*

64   **after it was used to cure**: D. Podolsky, *Cures out of Chaos: How Unexpected Discoveries Led to Breakthroughs in Medicine and Health* (Abingdon, UK: Routledge, 1998).

64   **sweet liquid**: Hilts, *Protecting America's Health.*

64   **shipped the elixir**: B. Martin, *Elixir: The American Tragedy of a Deadly Drug* (Lancaster, PA: Barkberry Press, 2014).

64   **doctors and patients had no idea**: Ibid.

64   **largest sum**: Hilts, *Protecting America's Health.*

65   **increasing federal authority**: Ibid.

65   **had not even existed**: Rosen, *Miracle Cure.*

65   **expanded**: Blum, *The Poison Squad.*

65   **asked to review**: G. W. Mellin and M. Katzenstein, "The Saga of Thalidomide: Neuropathy to Embryopathy, with Case Reports of Congenital Anomalies," *New England Journal of Medicine* 267 (1962): 1238–44.

65   **regulatory documents**: L. Bren, "Frances Oldham Kelsey: FDA Medical Reviewer Leaves Her Mark on History," *FDA Consumer* 35, no. 2 (2001): 24–29.

65   **phocomelia**: H. B. Taussig, "A Study of the German Outbreak of Phocomelia: The Thalidomide Syndrome," *Journal of the American Medical Association* 180 (1962): 1106–14.

65   **publicly attacked her**: S. Scheindlin, "The Courage of One's Convictions: The Due Diligence of Frances Oldham Kelsey at the FDA," *Molecular Interventions* 11, no. 1 (2011): 3–9.

66   **her careful work**: Ibid.

66   **feted her**: Bren, "Frances Oldham Kelsey," 24–29.

66   **until she was ninety**: G. Watts, "Frances Oldham Kelsey," *Lancet* 386, no. 10001 (2015): 1334.

66   **scope of this work**: Hilts, *Protecting America's Health.*

67   **If a company wants to price gouge**: G. M. Halpenny, "High Drug Prices Hurt Everyone," *ACS Medicinal Chemistry Letters* 7, no. 6 (2016): 544–46.

67   **ethical obligation**: D. Crow, "Pharma Chief Defends 400% Drug Price Rise as a 'Moral Requirement,'" *Financial Times*, September 11, 2018.

68   **"breakthrough therapy"**: J. Corrigan-Curay, A. E. McKee, and P. Stein, "Breakthrough-Therapy Designation—An FDA Perspective," *New England Journal of Medicine* 378, no. 15 (2018): 1457–58.

68   **lives in danger**: M. M. McLaughlin et al., "Developing a Method for Reporting Patient Harm Due to Antimicrobial Shortages," *Journal of Infectious Diseases and Therapy* 3, no. 2 (2014): 349–55.

68   **The price of doxycycline**: A. V. Gundlapalli et al., "Antimicrobial Agent Shortages: The New Norm for Infectious Diseases Physicians," *Open Forum Infectious Diseases* 5, no. 4 (2018): ofy068.

69   **shortages of antibiotics**: F. Quadri et al., "Antibacterial Drug Shortages from 2001 to 2013: Implications for Clinical Practice," *Clinical Infectious Diseases* 60, no. 12 (2015): 1737–42.

69   **"market failure"**: K. Guimaraes, "Why Is the World Suffering from a Penicillin Shortage?," Al Jazeera online, last modified May 21, 2017, www.aljazeera.com /indepth/features/2017/05/world-suffering-penicillin-shortage-17051707590 2840.html.

70   **entirely preventable**: D. A. Watkins et al., "Global, Regional, and National Burden of Rheumatic Heart Disease, 1990–2015," *New England Journal of Medicine* 377, no. 8 (2017): 713–22.

CHAPTER 9: Backwater

71   **have to vet**: H. W. Boucher et al., "White Paper: Developing Antimicrobial Drugs for Resistant Pathogens, Narrow-Spectrum Indications, and Unmet Needs," *Journal of Infectious Diseases* 216, no. 2 (2017): 228–36.

73   **seeping into the community**: F. D. Lowy, "Mapping the Distribution of Invasive *Staphylococcus Aureus* Across Europe," *PLOS (Public Library of Scence) Medicine* 7, no. 1 (2010): e1000205.

73   **effective against MRSA**: H. W. Boucher, G. H. Talbot, and M. W. Dunne, "Dalba-vancin or Oritavancin for Skin Infections," *New England Journal of Medicine* 371, no. 12 (2014): 1161–62.

75   **tribal sovereign immunity**: O. Dyer, "Allergan Transfers Restasis Patent to Mohawk Tribe to Deter Challenges from Generics," *British Medical Journal* 358 (2017): j4280.

78   **orphaned**: B. Shearer and B. Shearer, eds., *Notable Women in the Life Sciences: A Biographical Dictionary* (Westport, CT: Greenwood Press, 1996).

78   **bouncing around**: A. Espinell-Ingroff, *Medical Mycology in the United States: A His-torical Analysis (1894–1996)* (Dordrecht, Neth.: Springer Science and Business Me-dia, 2003).

78   **moved to New York**: Shearer and Shearer, *Notable Women.*

78   **served in the army**: Ibid.

78   **Hazen was tasked**: Espinell-Ingroff, *Medical Mycology.*

78   **mailed her samples**: Shearer and Shearer, *Notable Women.*

78   **blistering pace**: Espinell-Ingroff, *Medical Mycology.*

78   **after years of searching, one worked**: W. Dismukes, P. Pappas, and J. Sobel, *Clinical Mycology* (New York: Oxford University Press, 2003).

79   **Hazen's friend**: Espinell-Ingroff, *Medical Mycology.*

79   **named it after her friend**: E. Hazen and R. Brown, "Fungicidin, an Antibiotic Pro-duced by a Soil Actinomycete," *Proceedings of the Society for Experimental Biology and Medicine* 76, no. 1 (1951): 93–97.

79   **briefly, famous**: W. S. Bacon, "Elizabeth Lee Hazen, 1885–1975," *Mycologia* 68, no. 5 (1976): 961–69.

79   **funded more research**: E. Hazen and R. Brown, "Nystatin," *Annals of the New York Academy of Sciences* 89 (1960): 258–66.

79   **continued to collaborate**: R. Brown, E. L. Hazen, and A. Mason, "Effect of Fungici-din (Nystatin) in Mice Injected with Lethal Mixtures of Aureomycin and *Candida albicans*," *Science* 117, no. 3048 (1953): 609–10.

79   **penicillin has saved**: Brown, *Penicillin Man.*

80   **pricing system**: C. Årdal et al., "Insights into Early Stage of Antibiotic Development in Small- and Medium-sized Enterprises: A Survey of Targets, Costs, and Durations," *Journal of Pharmaceutical Policy Practice* 11 (2018): 8.

80   **an options market**: M. Savic and C. Årdal, "A Grant Framework as a Push Incentive to Stimulate Research and Development of New Antibiotics," *Journal of Law, Medicine & Ethics* 46, no. 1 supp. (2018): 9–24.

CHAPTER 10: Ruth

85   **gone into effect**: C. Eby, *Hungary at War: Civilians and Soldiers in World War II* (University Park, PA: Penn State University Press, 2007).

86   **an uneasy alliance**: Ibid.

CHAPTER 11: George

93   **screech and snarl of airplanes**: J. Diamond, *New Guinea: The Allied Jungle Campaign in World War II* (Mechanicsburg, PA: Stackpole Books, 2015).

94   **a tactical advantage on the ground**: J. Duffy, *War at the End of the World: Douglas MacArthur and the Forgotten Fight for New Guinea, 1942–1945* (New York: NAL Caliber, 2016).

94   **a common adversary**: Diamond, *New Guinea*.

CHAPTER 12: Mississippi Mud

99   **compound 05865**: R. S. Griffith, "Vancomycin Use—An Historical Review," *Journal of Antimicrobial Chemotherapy* 14 (1984): supp. D: 1–5.

99   **"Mississippi Mud"**: L. S. Elting et al., "Mississippi Mud in the 1990s: Risks and Outcomes of Vancomycin-Associated Toxicity in General Oncology Practice," *Cancer* 83, no. 12 (1998): 2597–607.

CHAPTER 13: Soren

106   **overdoses related to prescription opiates**: G. Comerci, J. Katzman, and D. Duhigg, "Controlling the Swing of the Opioid Pendulum," *New England Journal of Medicine* 378, no. 8 (2018): 691–93.

CHAPTER 14: Duty

111   **all kinds of malignancies**: Centers for Disease Control and Prevention, "World Trade Center Health Program: Addition of Certain Types of Cancer to the List of

WTC-Related Health Conditions—Final Rule," *Federal Register* 77, no. 177 (2012): 56138–68.

CHAPTER 15: Remy

117   **Tom and I had written about**: M. McCarthy et al., "Mold Infections of the Central Nervous System. *New England Journal of Medicine* 371, no. 2 (2014): 150–60.
118   **puppies**: M. P. Montgomery et al., "Multidrug-Resistant *Campylobacter jejuni* Outbreak Linked to Puppy Exposure—United States, 2016–2018," *Morbidity and Mortality Weekly Report (MMWR)* 67, no. 37 (2018): 1032–35.

CHAPTER 16: A Quiet Revolution

120   **figured out how to harness**: T. F. Gajewski, "Fast Forward—Neoadjuvant Cancer Immunotherapy," *New England Journal of Medicine* 378, no. 21 (2018): 2034–35.
124   **doctors had never seen before**: D. Yong et al., "Characterization of a New Metallo-Beta-Lactamase Gene, bla(NDM-1), and a Novel Erythromycin Esterase Gene Carried on a Unique Genetic Structure in *Klebsiella pneumoniae* Sequence Type 14 from India," *Antimicrobial Agents and Chemotherapy* 53, no. 12 (2009): 5046–54.
124   **where it might spread**: A. Marra, "NDM-1: A Local Clone Emerges with Worldwide Aspirations," *Future Microbiology* 6, no. 2 (2011): 137–41.
124   **concluded**: D. Yong et al., "Characterization of a New Metallo-Beta-Lactamase Gene," 5046–54.

CHAPTER 17: Decision Points

128   **"an almost Victorian attitude"**: K. A. Sepkowitz, "Finland, Weinstein, and the Birth of Antibiotic Regret," *New England Journal of Medicine* 367, no. 2 (2012): 102–3.

CHAPTER 19: Garden State

136   **discovered in the ear**: K. Satoh et al., "*Candida auris* sp. nov., a Novel Ascomycetous Yeast Isolated from the External Ear Canal of an Inpatient in a Japanese Hospital," *Microbiology and Immunology* 53, no. 1 (2009): 41–44.

CHAPTER 20: Trojan Horses

141   **figured out a way**: M. Abbas, M. Paul, and A. Huttner, "New and Improved? A Review of Novel Antibiotics for Gram-Positive Bacteria," *Clinical Microbiology and Infection* 23, no. 10 (2017): 697–703.

142 **made other drugs more powerful**: O. Lomovskaya et al., "Vaborbactam: Spectrum of Beta-Lactamase Inhibition and Impact of Resistance Mechanisms on Activity in *Enterobacteriaceae*," *Antimicrobial Agents and Chemotherapy* 61, no. 11 (2017): e01443-17.

142 **Trojan horse approach**: A. Górska, A. Sloderbach, and M. P. Marszałł, "Siderophore-Drug Complexes: Potential Medicinal Applications of the 'Trojan Horse' Strategy," *Trends in Pharmacological Sciences* 35, no. 9 (2014): 442–49.

143 **price hike**: J. D. Alpern, "Trends in Pricing and Generic Competition Within the Oral Antibiotic Drug Market in the United States," *Clinical Infectious Diseases* 65, no. 11 (2017): 1848–52.

CHAPTER 21: The Rockefellers

147 **hawking potions and elixirs**: R. Chernow, *Titan: The Life of John D. Rockefeller, Sr.* (New York: Vintage, 2004).

148 **donated to charity**: Ibid.

148 **solitude of a rocking chair**: Ibid.

148 **worked initially as a middleman**: Ibid.

148 **edged out his partners**: Ibid.

149 **had an idea**: Ibid.

149 **one unique request**: Ibid.

149 **meningitis outbreak**: S. Flexner and J. W. Jobling, "An Analysis of Four Hundred Cases of Epidemic Meningitis Treated with the Anti-Meningitis Serum," *Journal of Experimental Medicine* 10, no. 5 (1908): 690–733.

150 **visited his East River campus just once**: Chernow, *Titan*.

CHAPTER 23: Breakthrough

157 **first purified the protein**: V. A. Fischetti, E. C. Gotschlich, and A. W. Bernheimer, "Purification and Physical Properties of Group C Streptococcal Phage-Associated Lysin," *Journal of Experimental Medicine* 133, no. 5 (1971): 1105–17.

CHAPTER 24: Anthrax

162 **brought him to an emergency room**: L. M. Bush et al., Index Case of Fatal Inhalational Anthrax Due to Bioterrorism in the United States," *New England Journal of Medicine* 345, no. 22 (2001): 1607–10.

163 **fever and confusion**: Ibid.

163 **seizure**: Ibid.

163 **temperature rose**: Ibid.

164 **Daschle announced that anthrax had been found**: D. B. Jernigan et al., "Investigation of Bioterrorism-Related Anthrax, United States, 2001: Epidemiologic Findings," *Emerging Infectious Diseases* 8, no. 10 (2002): 1019–28.

165   **heat wave**: M. G. Walsh, A. W. de Smalen, S. M. Mor, "Climatic Influence on Anthrax Suitability in Warming Northern Latitudes," *Scientific Reports* 8, no. 1 (2018): 9269.

165   **it reported about his work**: R. Schuch, D. Nelson, and V. A. Fischetti, "A Bacteriolytic Agent That Detects and Kills *Bacillus anthracis*," *Nature* 418, no. 6900 (2002): 884–89.

Chapter 27: Mantra

184   **had been described in *Vanity Fair***: T. J. English, "A Brief History of Mayoral–N.Y.P.D. Dysfunction," *Vanity Fair*, January 29, 2015.

Chapter 28: Obstacles

189   **the net present value of a new antibiotic**: D. M. Brogan and E. Mossialos, "A Critical Analysis of the Review on Antimicrobial Resistance Report and the Infectious Disease Financing Facility," *Global Health* 12 (2016): 8.

Chapter 31: Persuasion

201   **ProPublica found**: C. Chen and R. Wong, "Black Patients Miss Out on Promising Cancer Drugs," ProPublica, last modified September 19, 2018, www.propublica.org/article/black-patients-miss-out-on-promising-cancer-drugs.

Chapter 32: The Rollout

207   **killing hundreds of thousands**: D. L. Hammarlöf et al., "Role of a Single Noncoding Nucleotide in the Evolution of an Epidemic African Clade of *Salmonella*," *Proceedings of the National Academy of Sciences of the United States of America* 115, no. 11 (2018): E2614–23.

207   **spreading through Pakistan**: J. R. Andrews et al., "Extensively Drug-Resistant Typhoid—Are Conjugate Vaccines Arriving Just in Time?," *New England Journal of Medicine* 379, no. 16 (2018): 1493–95.

Chapter 33: Investments

210   **first detected**: R. Leclercq et al., "Transferable Vancomycin and Teicoplanin Resistance in *Enterococcus faecium*," *Antimicrobial Agents and Chemotherapy* 33, no. 1 (1989): 10–15.

Chapter 34: Into the Haystack

216 **"malacidins"**: B. M. Hover et al., "Culture-Independent Discovery of the Malacidins as Calcium-Dependent Antibiotics with Activity Against Multidrug-Resistant Gram-Positive Pathogens," *Nature Microbiology* 3, no. 4 (2018): 415–22.

Chapter 37: Searching

228 **published the work**: T. Kaeberlein, K. Lewis, and S. S. Epstein, "Isolating 'Uncultivable' Microorganisms in Pure Culture in a Simulated Natural Environment," *Science* 296, no. 5570 (2002): 1127–29.

228 **announced their findings**: L. L. Ling et al., "A New Antibiotic Kills Pathogens Without Detectable Resistance," *Nature* 517, no. 7535 (2015): 455–59.

Index